ELECTRONIC VARIABLE SPEED DRIVES

Second Edition

Michael E. Brumbach

DELMAR

™

THOMSON LEARNING

Australia Canada Mexico Singapore Spain United Kingdom United States

Electronic Variable Speed Drives 2E

Michael E. Brumbach

Business Unit Director:
Alar Elken

Executive Editor:
Sandy Clark

Acquisitions Editor:
Mark Huth

Development:
Dawn Daugherty

Executive Marketing Manager:
Maura Theriault

Marketing Coordinator:
Brian McGrath

Executive Production Manager:
Mary Ellen Black

Senior Production Coordinator:
Toni Hansen

Art/Design Coordinator:
Rachel Baker

Library of Congress Cataloging-in-Publication Data:

Brumbach, Michael E., 1955-
  Electronic variable speed drives / Michael E. Brumbach.—2nd ed.
      p. cm.
  Includes index.
  ISBN 0-7668-2839-5 (pbk.)
  1. Electric driving, Variable speed. I. Title.
TK4058 .B78 2001
621.46—dc21

                                2001032552

NOTICE TO THE READER

*"Thinking always of trying to do more
brings a state of mind in which
nothing seems impossible."*
Henry Ford

for
Jen, Kathy, Mom, and in loving memory of Dad

TABLE OF CONTENTS

PREFACE

Electronic Variable Speed Drives 2/E is intended for individuals who are in, or plan to enter, the maintenance field. It is targeted primarily toward students in two-year technical schools, but it should also be suitable for training in an industrial environment or as an introductory course at four-year institutions.

This text has been written with its intended audience in mind. Theories of operation have been greatly simplified. Very few formulas are presented, and those that are used are very simple. It has been my experience that maintenance personnel are some of the most intelligent people on the planet. Their nature is to get quickly to the heart and soul of the matter at hand. This text has been designed to do just that by providing in straightforward language the material that the maintenance technician needs to get the job done!

This book was designed to support a sixteen-week course on electronic variable speed drives. Therefore, it is assumed that students already understand DC motors, AC motors, and solid-state devices. Should a review of these components be necessary, additional material on these areas and several manufacturers' schematics of actual drives have been included in the Appendices. A glossary following the Appendices will help the student understand unfamiliar terminology.

Basically, this text is divided into two sections: eight chapters on DC drives and six chapters on AC drives. Each chapter and section begins with a general overview. Review questions keyed to explicitly stated learning objectives conclude the chapters. The two longest chapters (8 and 14) cover maintenance and troubleshooting, in keeping with the responsibilities of the target audience.

The second edition of *Electronic Variable Speed Drives* boasts several significant improvements over the first edition. Perhaps the most predominant change is the addition of numerous waveforms at appropriate points within the various drive schematics. The schematics shown in Chapters 2, 3, and 11 demonstrate this addition. Also, in these same chapters, improved drawings showing the timing of the various waveforms and the associated conduction times of the power semiconductors have been added. This should aid the student in better understanding the operation of a DC or AC drive. Chapters 8 and 14, which deal with troubleshooting, have had additional material added that covers power semiconductor modules. Schematics of the modules and testing with an ohmmeter are discussed. Flux vector drives can be found in Chapter 13. In the first edition, flux vector drives were found in Appendix F. This section has been moved to its own chapter to give the topic more prominence. Finally, several errors and inconsistencies have been corrected.

ACKNOWLEDGMENTS

I must begin with my sincere thanks to Olin Marth, who got me interested in electricity and electronics back in 1973 at Muhlenberg Senior High School. His introduction to Ohm's law and soldering have served me well these many years. (However, I still think I deserved an "A" that first quarter!)

Next I would like to thank all of the fine instructors and staff at Lincoln Technical Institute in Allentown, PA, in particular, the late Alex Bosico, the retired Norman Lee, and the past Director, Robert Milot. (Remember class 240-B?)

Thanks are also due to my students at York Technical College in Rock Hill, SC, who, on finding out that I was working on a textbook, took great interest in supplying me with a wealth of documents and industry contacts.

I also want to express my gratitude to Dan Stroud of Carotron, in Heath Springs, SC, for his tremendous help in the form of manufacturers' schematics, and clarification of certain matters of theory. Matthew O'Kane of Control Techniques in Grand Island, NY, and Bruce Wydell of Danfoss Drives in Rockford, IL, were also very generous in supplying information for this project.

I also wish to thank the folks at Delmar, Mark Huth, my editor, and Dawn Daugherty, my assistant editor.

I would be remiss if I did not also acknowledge some very important people from York Technical College. First, the other woman in my life, Cree Stout. Cree has been like a sister to me, helping, guiding, coaching, and listening tirelessly. Her unfailingly constructive criticisms have been of invaluable assistance on countless occasions. The other person, equally important and unquestionably unique, is my mentor and former department manager, Charlie Peek, one of the most unselfish individuals I have ever had the pleasure of knowing. Without his encouragement, I would not be where I am today. I owe him a great debt of gratitude—more than he is willing to accept—and I will always look forward to our customary five o'clock "discussions." I also want to thank Al Streeter. Al has been a great source of tips and insights into the theory behind drives. He has also provided gentle, constructive criticism as well as numerous suggestions on how the second edition could be better. Al is a great instructor and I am fortunate to have him on my team. Lastly, I must thank my "little brother," JC Clade. I had the pleasure of meeting JC when he was one of my students a few years back. JC so impressed me, that when a teaching position opened at the college, he was first on my list. I have never regretted hiring JC and I hope I never lose him. He has been an inspiration to his students and to me. Truth be known, he keeps me on my toes! JC has played a big part in this revision. His suggestions and willingness to help me see how to explain things more clearly will help make the second edition better. I am very fortunate to know and work with JC, but I am even more fortunate to be able to call him my friend. I wish him all the best in life.

Finally, to my parents, who have been with me all the way, a heartfelt "Thank you!" Whenever I needed you, you were there, but I do miss you, Dad. Words cannot express my appreciation. I also hope that my daughter, Jen, realizes how important she has been as a source of inspiration and motivation throughout this project. (I guess we'll be able to pay that college tuition now!) I wish her a happy and successful future. I love you all so very much.

And lastly, to my wife, Kathy, thanks for all of your patience and understanding each time my mood sagged under the weight of meeting yet another deadline. I guess third time is the charm! I love you more than you can know. Now, I'll take you shopping!

The author and Delmar gratefully acknowledge the time and effort of those who were on the review panel for this project. Their comments and suggestions proved most helpful.

Thomas Pickren
Albany Technical Institute
Albany, GA

Dick Dierks
Fox Valley Technical College
Appleton, WI

Will Parker
Angelina College
Lufkin, TX

INTRODUCTION

After meeting with industry representatives, my colleagues in the Industrial Maintenance Department and I determined that the need existed for a course in DC and AC electronic variable speed drives as part of our Industrial Electricity/Electronics curriculum at York Technical College. Having identified a suitable course, we began to search for a suitable text and could not find one. Most devoted only one or two chapters to electronic variable speed drives—far too little material to occupy a sixteen-week semester. Necessity being the mother of invention, I began to write my own text.

In doing so, I have tried to maintain a course level that is suitable for my curriculum and for two-year technical schools in general. I deliberately avoided developing a text oriented toward engineers. Our students will have careers in the Industrial Maintenance areas and have no need nor use for in-depth design theory, formulas, or circuit analysis. What they need is a no-nonsense approach to the nitty-gritty, basic theories of electronic drive operation, written in a language that they can understand. This is the objective I tried to accomplish as I wrote *Electronic Variable Speed Drives*.

The chapters have purposely been kept short. Very few formulas are used. Circuit explanations are very simple and not every component is explained. The longest chapters deal with troubleshooting and maintenance.

In order to enhance this text, I would like to suggest some additional resources that instructors may find helpful. For video tapes, we have found a three-part series on solid-state motor controls from Tel-A-Train to be quite useful. The material covered includes DC drives, AC drives, and servo/stepper drives. In addition, Tel-A-Train provides computerized testing software and workbooks. Tel-A-Train can be reached at: P.O. Box 4752, 309 North Market Street, Chattanooga, TN 37405. Their telephone number is (800) 251-6018. Another noteworthy resource is Danfoss Drives interactive software. This software allows the student to review the theory of motor operation, connect a drive, program the drive for various acceleration and deceleration rates, and so on. Danfoss Drives can be reached at: 2995 Eastrock Drive, Rockford, IL 61109. Their telephone number is (800) 432-6367.

Chapter 1

DC Drive Fundamentals

OBJECTIVES

After completing this chapter, you will be able to:

- Discuss the general purpose of a DC drive.

- Discuss the general operating principle of a DC drive.

- List four parameters of a DC motor that a DC drive can control.

- Define the following terms: torque, counter EMF, command signal, error signal, feedback signal, and regulation.

DC motors have been used in industry for many years. Typically, these motors have been operated at full speed or slightly below. When necessary, the speed of the motor may be varied, usually by adding series resistance. Although this practice works well and is easy to accomplish, it is inefficient, wasting a large amount of energy in the form of heat given off by the resistors.

Electronic variable speed drives are also used to vary the speed of DC motors with greater **efficiency** and more precise speed control than resistors offer. Subsequent chapters will present other ways to control DC motor speed. In this chapter we will discuss how DC electronic variable speed drives such as the one shown at the top of the next page in Figure 1-1 perform this task.

What Is the Structure of an Electronic Variable Speed Drive?

An **electronic variable speed drive** has two basic sections: the control section and the power section. The control section governs or controls the power section while the power section supplies controlled power to the DC motor. (See Figure 1-2, page 3.)

The control section allows us to control not only the motor's speed but also its torque (Recall that **torque** is the turning force that a motor produces.) Motor

1

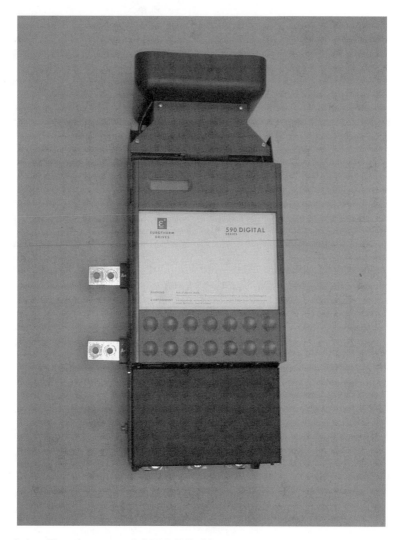

Figure 1-1: *Eurotherm model 590 DC drive.*

speed and torque control can be accomplished by one of two methods: We can vary either the voltage to the armature of the DC motor or the current to the **field.** When we vary the armature voltage, the motor produces full torque, but the speed is varied. However, if the field current is varied both the motor speed **and** the torque will vary.

Because of the need to vary the armature voltage or the field current, a separately excited DC motor, which allows very precise control over speed and torque, is the most commonly used type of motor.

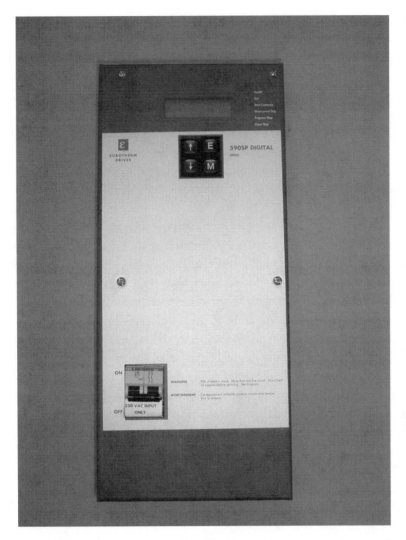

Figure 1-2: *Eurotherm model 590SP DC drive with programming panel.*

How Does an Electronic Variable Speed Drive Work?

Attaining precise control over motor speed and torque requires a means of evaluating the motor's performance and automatically compensating for any variations from the desired levels. The control section of a DC drive uses three types of signals to evaluate motor performance: the command, feedback, and error signals.

A **command signal,** sometimes called the **set point** or **reference signal,** is programmed into the DC drive and sets the desired operating speed of a DC motor. While the motor is operating, a **feedback signal** from the motor indicates the motor's performance. The feedback signal can originate either from the counter electromotive force produced by the motor or from a tachometer-generator or **encoder** mounted on the motor's shaft. **Counter electromotive force (CEMF)** is the voltage produced in the motor's rotating armature, which cuts the magnetic lines of force in the field as it revolves. This armature voltage is called counter EMF because it opposes the applied voltage from the electronic variable speed drive. As a result of this opposing voltage, the amount of armature current will be limited. A difference exists between the control signal and the feedback signal. This difference is the **error signal.** The electronic variable speed drive's controller automatically adjusts the motor's performance until the error signal is reduced practically to zero. This function is an ongoing process.

A controller's **regulation** determines how well it responds to changes in motor performance and is usually expressed as a percentage. Different types of feedback signals result in variable degrees of regulation. For instance, controllers using counter EMF feedback, also called **armature voltage feedback,** typically have a regulation of 5% to 8%. This means that the speed of a DC motor set to operate at 1800 RPM can vary from 1944 RPM to 1656 RPM (1800 RPM ± 8%). On the other hand, when shaft-coupled encoders are used to provide the feedback signal, regulation is much tighter. A typical shaft-coupled encoder produces a regulation of 0.01%. Thus the speed variation of the same motor operating at 1800 RPM would drop to between 1799.92 RPM and 1800.18 RPM (1800 RPM ± 0.01%)!

In addition to managing motor speed and torque, the control section of a DC drive determines the direction of motor rotation and controls motor braking as we will see in the following chapters.

Review Questions

1. Name the four parameters of a DC motor that a DC drive controls.

2. Name at least two sources of the feedback signal.

3. Explain the importance of the feedback signal. Why is it needed?

4. What does the term "error signal" mean?

5. **True or False?** A higher percentage of regulation is better than a lower percentage of regulation. Explain.

6. What is meant by counter EMF (CEMF)?

7. **True or False?** Counter EMF is undesirable. Explain your answer.

8. Torque is best defined as

   a. an opposing voltage produced in a motor's armature.

   b. the difference between the actual speed and the desired speed of a motor.

   c. an undesirable phenomenon, and steps should be taken to eliminate it.

   d. the turning force produced by a motor.

Chapter 2

Switching Amplifier Field Current Controllers

OBJECTIVES

After completing this chapter, you will be able to:

- Explain the concepts of open-loop and closed-loop control.

- Discuss the theory and operation of a switching amplifier field current controller.

- Define the following terms: tachometer-generator, feedback signal, reference or command signal, summing point, error signal, and comparator.

In this chapter, we examine the switching amplifier field current controller and introduce the concepts of feedback and closed-loop control. We also discuss how the speed of a DC shunt-wound motor can be held constant under varying load conditions.

Before you can understand how a switching amplifier field current controller works, you must understand what is meant by open-loop control and closed-loop control. In **open-loop control,** also called **manual control,** any variations in motor speed must be compensated by a manual adjustment. As you can imagine, paying someone to monitor and adjust motor speed constantly under varying load conditions would be very costly and inefficient. Therefore, open-loop control is limited to applications where the motor load is fairly constant.

When the motor load varies considerably or frequently, closed-loop control is used. **Closed-loop control,** also called **automatic control,** uses feedback information to monitor the performance of a motor. The information that is fed back automatically causes the control circuit to adjust the motor speed to varying load conditions. In the following section we will consider how closed-loop control is used in a switching amplifier field current controller.

What Is a Switching Amplifier Field Current Controller?

Basically, a **switching amplifier field current controller,** as shown in Figure 2-1, receives and responds to a feedback signal from a DC motor that provides information about the speed of the motor. The controller first compares the feedback signal to a **reference signal.** Depending on the result of this comparison, the controller automatically increases or decreases the motor speed until it reaches the level indicated by the reference signal. To accomplish this adjustment, the controller switches the shunt field current on and off at varying rates.

How Does the Switching Amplifier Field Current Controller Do This?

To understand the following detailed explanation, refer to Figure 2-1. We begin with the feedback device.

In the upper right corner of Figure 2-1 is a schematic view of **tachometer-generator** (tach. gen.), a DC generator attached to the DC shunt motor shaft. As the DC shunt motor turns, the DC tachometer-generator also turns and produces a positive DC voltage as a result. The faster the DC shunt motor turns, the more positive DC voltage the DC tachometer-generator produces. Thus the output of the DC tachometer-generator is proportional to the speed of the DC shunt motor.

Next, we will consider the feedback section, shown in detail in Figure 2-2 (see page 10). The positive voltage produced by the DC tachometer-generator and fed through R1 into buffer amplifier U1 is the feedback signal. Resistor R1 is known as a scaling resistor. The value of R1 is adjusted to provide the proper amount of DC voltage from the tachometer-generator for a given RPM. The output of U1 is fed to the inverting input of U2, which inverts the polarity of the tachometer-generator voltage from positive to negative. The resulting negative voltage is then fed through R7 to the noninverting input of U3 in the preamplifier section. Note that one end of R8 is connected to the noninverting input of U3 and the other end of R8 is connected to variable resistor R6. Resistor R6 allows us to set the reference voltage level, which determines at what speed the DC shunt motor will run. This reference voltage may also be called the **command signal** or **set point.** The reference voltage (positive) and the tachometer-generator voltage (negative) are added together at the junction of R7 and R8, called the **summing**

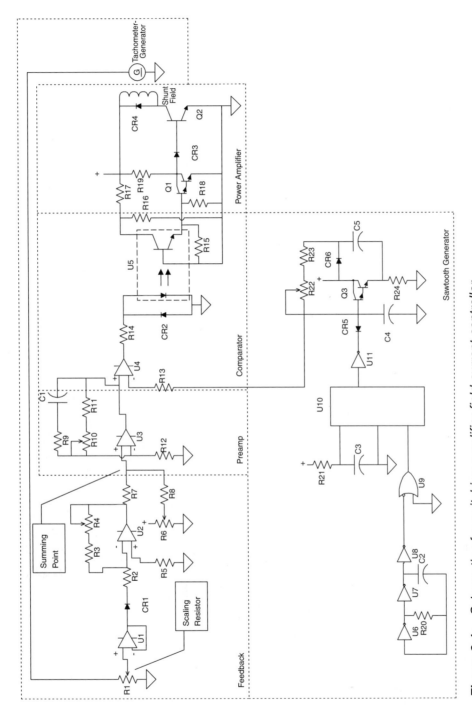

Figure 2-1: Schematic of a switching amplifier field current controller.

9

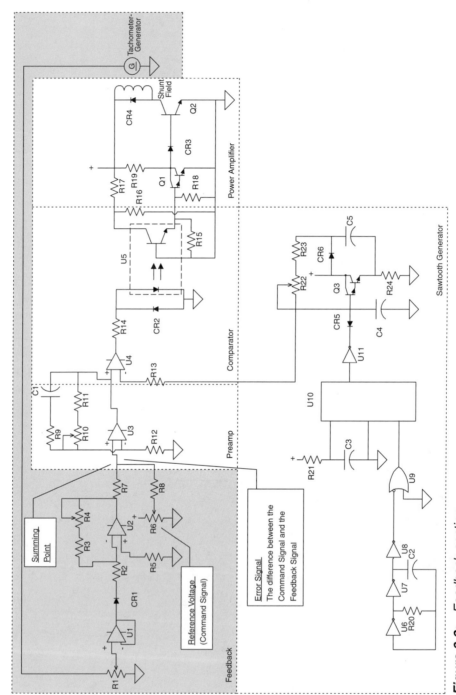

Figure 2-2: Feedback section.

10

point. The difference of these two voltages, which appears at the noninverting input of U3, is the **error signal.**

To understand how this process works, assume that we wish the DC shunt motor to turn at 1800 RPM. We find that R6 must be set to 10 volts, the reference voltage required to attain a motor speed of 1800 RPM. Assume also that at a speed of 1800 RPM, the DC tachometer-generator produces a positive output of 5 volts. This value will be inverted by U2. Thus at the junction of R7 and R8 we will have a positive reference voltage of 10 volts and a negative DC tachometer-generator feedback voltage of 5 volts. The sum of these two voltages is a positive error voltage of 5 volts $(10 + (-5) = 5)$.

If the speed of the DC shunt motor decreases below the reference speed, the tachometer-generator produces less positive DC voltage. Now assume that the DC tachometer-generator produces a positive voltage of 3 volts DC, which results in a negative voltage of 3 volts at R7 from the output of U2. Because the reference voltage does not change, we still have a positive voltage of 10 volts at R8 from R6. The sum of these two voltages at the junction of R7 and R8 is a positive error voltage of 7 volts $(10 + (-3) = 7)$. Therefore we can conclude that decreases in motor speed produce higher positive voltage at the noninverting input of U3. Likewise, increases in motor speed produce lower positive voltage at the noninverting input of U3.

What Does the Preamplifier Stage Do?

Figure 2-3 illustrates U3, a preamplifier stage which takes the voltage at its noninverting input and provides an amplified positive output voltage of the proper level for the comparator to work with. Before examining the comparator stage, we will first look at the sawtooth generator stage.

What Is the Purpose of the Sawtooth Generator Stage?

Inverters U6 and U7, together with R20 and C2, form an oscillator circuit as shown in Figure 2-4 (see page 13). Assume that the circuit's frequency of oscillation is 3 kHz. This oscillator circuit produces a rectangular pulse that is applied to U10, which is a pulse shaping circuit. The output of U10, along with that of Q3, produces a 3 kHz sawtooth ramp voltage that is applied to the inverting input of U4 in the comparator stage.

What Is a Comparator?

In the comparator, shown in Figure 2-5 (see page 14), the noninverting input of U4 receives a positive voltage from the output of the preamplifier stage.

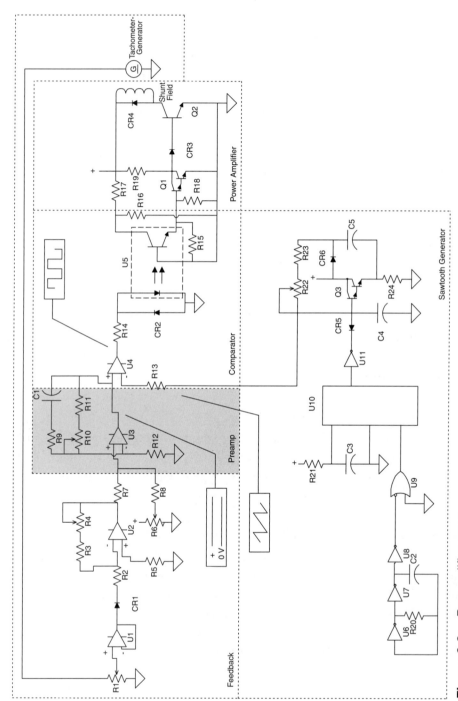

Figure 2-3: *Preamplifier stage.*

12

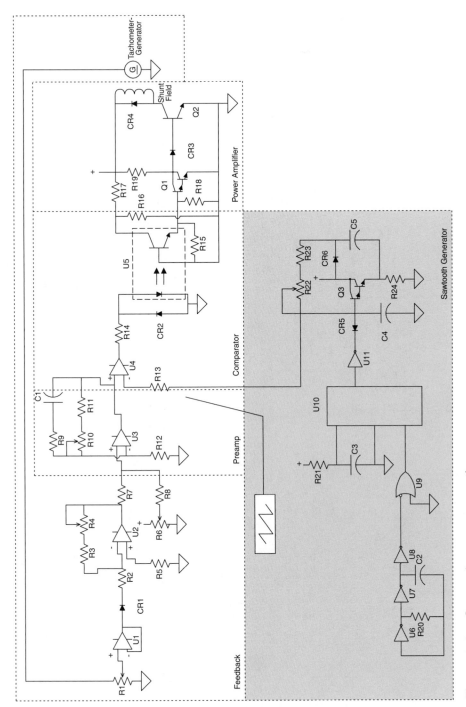

Figure 2-4: Sawtooth generator section.

13

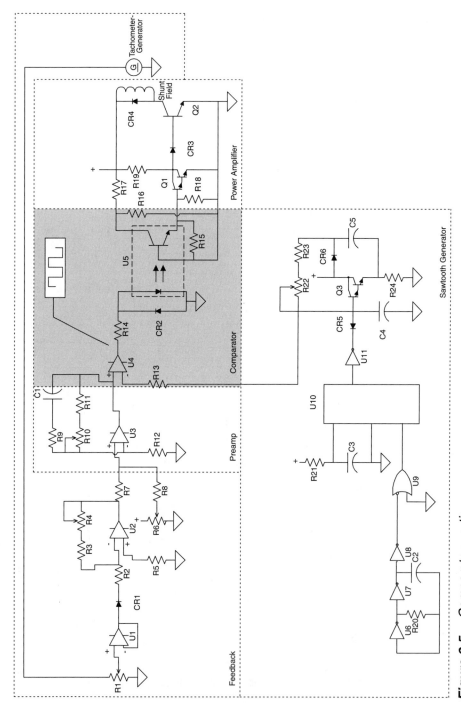

Figure 2-5: *Comparator section.*

Simultaneously, the inverting input of U4 receives a positive 3 kHz sawtooth ramp voltage from the sawtooth generator stage. This 3 kHz sawtooth voltage causes U4 to switch on and off. Whenever the noninverting input has a higher positive value than the inverting input from the sawtooth generator, U4 will produce an output voltage. Conversely, when the inverting input from the sawtooth generator has a higher positive value than the noninverting input, the output of U4 will be turned off, and a rectangular pulse will be produced at the output of U4 as a result. (See Figure 2-6.) The width of U4's output pulse is controlled by the reference voltage from R6. If R6 is adjusted for a higher DC reference voltage, then U3 will produce a higher positive voltage for the noninverting input of U4. In turn, the sawtooth ramp voltage must increase to a higher positive value at the inverting input of U4. Since the inverting input takes more time to attain a higher positive value than the noninverting input

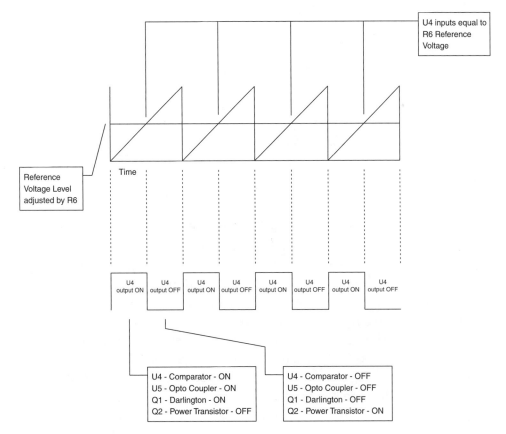

Figure 2-6:   *Output of the comparator section.*

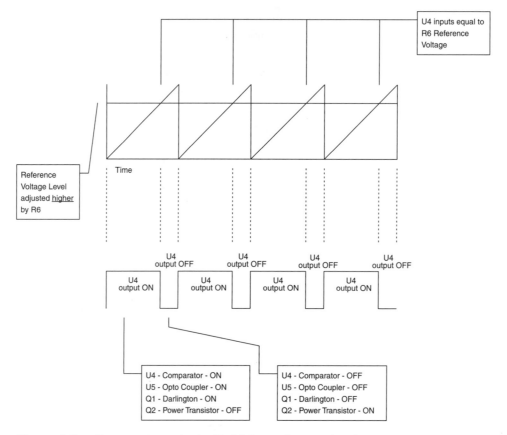

Figure 2-7: *Comparator output with higher reference level.*

does, the output pulse of U4 becomes wider. (See Figure 2-7.) Likewise, if R6 is adjusted for a lower DC reference voltage, U3 will produce a lower positive DC voltage at the noninverting input of U4. Consequently, the sawtooth ramp voltage does not need to increase to as high a positive value, and U4 turns off sooner. In this case U4's output pulse will be narrower. (See Figure 2-8, page 17.) This effect is called **pulse width modulation,** or **PWM.** The output pulse of U4 is applied to optocoupler (or optoisolator) U5. When the output of U4 is on, optocoupler U5 is on. When the output of U4 is off, optocoupler U5 is also off. Optocoupler U5 isolates the low-voltage logic circuits from the higher voltage power circuits in the power amplifier stage and also prevents electrical noise that may be induced onto the power circuits from entering the control section of the drive.

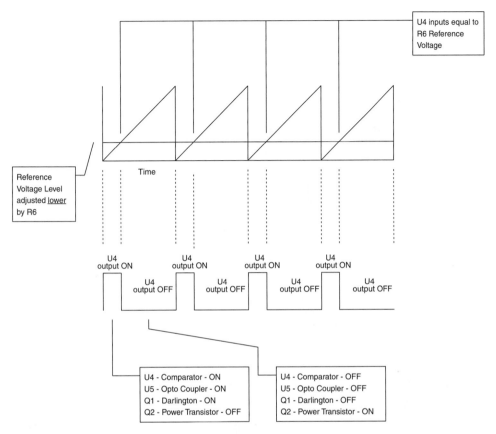

Figure 2-8: *Comparator output with lower reference level.*

What Is the Function of the Power Amplifier Stage?

Figure 2-9 (see page 18) illustrates how the output of optocoupler U5 is fed to the base of Q1 in the power amplifier stage. The power amplifier stage consists primarily of Darlington **transistor** Q1 and power transistor Q2. Transistor Q1 receives its input signal from optocoupler U5. Whenever U5 turns on, Q1 will conduct. Likewise, whenever U5 is turned off, Q1 will not conduct. The collector of Q1 drives the base of Q2. Whenever Q1 is conducting, Q2 is turned off, and no current flows through the shunt field of the DC shunt motor. As a result, the DC shunt motor speeds up. Conversely, whenever Q1 is turned off, Q2 does conduct a current, which flows through the shunt field of the DC shunt motor, causing the motor to slow down.

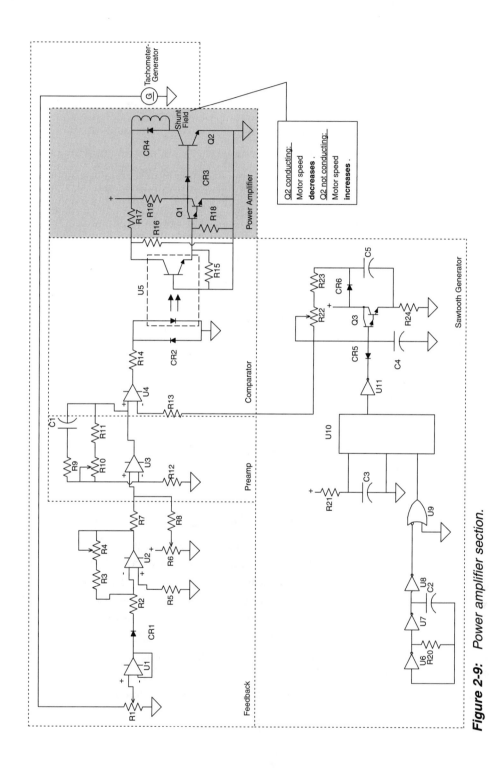

Figure 2-9: Power amplifier section.

The Closed-Loop Speed Control Process

How Does the Process Described in the Preceding Section Provide Closed-Loop Speed Control of a DC Shunt Motor?

Reconsider our earlier example of a motor set to a speed of 1800 RPM. Recall that we adjusted R6 for this speed and that this produced 10 volts DC at R8. Because the DC shunt motor is turning at 1800 RPM, the tachometer-generator produces a negative voltage of 5 volts DC at R7. Therefore at the junction of R7 and R8 we have a positive voltage of 5 volts DC. This voltage is applied to the noninverting input of the comparator, U4, and then compared with the sawtooth ramp voltage at the inverting input of U4. Assume that as a result of this comparison the output of U4 is turned on for 166 microseconds. Consequently, U5 and Q1 are also turned on for 166 microseconds. However, Q2 is turned off for 166 microseconds. Because 166 microseconds represents a 50% **duty cycle** at 3 kHz, the outputs of U4, U5, and Q1 will all be switched on and then off every 166 microseconds (% duty cycle = time of "on" pulse/time of one cycle). Consequently, Q2 will also be turned on and off every 166 microseconds. Therefore the current through the shunt field of the DC motor pulsates at an average value that keeps the motor turning at 1800 RPM.

What Happens If the Load on the DC Shunt Motor Is Increased?

Applying an increased load to the DC shunt motor slows the motor down. As the DC shunt motor turns more slowly, the tachometer-generator produces less output voltage. This reduction in voltage results in a lower negative voltage at R7. Suppose that the tachometer-generator voltage drops from 5 volts to 3 volts. The resulting voltage at the junction of R7 and R8 will be positive 7 volts $(10 + (-3) = 7)$. Consequently, a higher voltage (7 volts) is applied to the noninverting input of U4. When this voltage is compared to the sawtooth ramp voltage, the value of the sawtooth ramp voltage must increase to a higher value before U4 is turned off. Thus the output of U4 will be turned on for a longer period of time than it was previously. Assume that U4 is now turned on for 233 microseconds. This implies that the output of U4 will be turned off for a shorter period of time (100 microseconds). The longer U4 is turned on and Q2 is turned off, the longer will be the time during which no current flows through the shunt field of the DC shunt motor. Therefore the average value of the current flowing through the **shunt field** will be lower. Reducing the shunt field current causes the DC shunt motor to speed up. The reverse is true if the load on the DC shunt motor is lessened. In this way, an increase or decrease in load on a DC shunt motor is automatically compensated. The process by which this compensation is accomplished is closed-loop control, or automatic control.

Review Questions

1. Describe the operation of a tachometer-generator.

2. Is the voltage generated by a tachometer-generator proportional or inversely proportional to motor speed?

3. **True or False?** The summing point is that point in a circuit where the output from the sawtooth generator is joined with the output from the preamplifier.

4. Explain the function of the comparator stage.

5. An optocoupler:

    a. protects low voltage circuits from high voltage circuits.

    b. provides electrical noise immunity.

    c. is also called an optoisolator.

    d. a and b only.

    e. a, b, and c.

6. Explain what is meant by closed-loop control.

7. What is another name for closed-loop control?

8. If the load on a DC motor varies frequently, would you use a DC drive with open- or closed-loop control? Why?

Challenge Yourself

1. Look at Figure D-1 in Appendix D. If you were asked to connect this DC drive, to which terminals would you make the following connections?

    a. AC supply L1 to terminal _____.

    b. AC supply L2 to terminal _____.

    c. AC ground to terminal _____.

    d. Armature A1 to terminal _____.

    e. Armature A2 to terminal _____.

    f. Field F1 to terminal _____.

    g. Field F2 to terminal _____.

2. Explain what modification must be made to the drive in Exercise 1 if the operating voltage is 115 V rather than 230 V.

3. Explain the effect of this modification on the drive circuit.

4. Look at Figure D-2 in Appendix D, and answer the following questions.

    a. What is connected to terminals 1, 2, and 3? Explain the purpose of this device.

    b. Something remains to be connected to terminals 4 and 5. What is it? Where does it come from? What controls it? What does it do?

    c. If you are using a tachometer-generator for feedback, to what terminals must it be connected?

    d. What is the required output of the tachometer-generator?

Chapter 3

SCR Armature Voltage Controllers

OBJECTIVES

After completing this chapter, you will be able to:

■ Discuss the similarities between the SCR armature voltage controller and the switching amplifier field current controller.

■ Discuss the differences between the SCR armature voltage controller and the switching amplifier field current controller.

■ Discuss the theory and operation of an SCR armature voltage controller.

This chapter presents another method of DC motor speed control: **armature voltage control**. We will learn how this type of control is accomplished using closed-loop control and SCRs.

What Is an SCR Armature Voltage Controller?

An SCR armature voltage controller is another method of DC motor speed control. The speed of a DC motor may be varied by controlling either the shunt field current or the armature voltage. Armature voltage control is the type most commonly used in DC drives. In Chapter 2 we learned how to control the shunt field current. Now we will see that by controlling the firing of SCRs we can control the voltage of the armature of a DC motor. A typical SCR armature voltage controller is shown in Figure 3-1 on page 24.

How Can the Armature Voltage Be Controlled?

As you look at Figure 3-2, try to recognize some of its similarities in relation to Figure 2-3 (see page 12). Both units use the same type of circuit for closed-loop

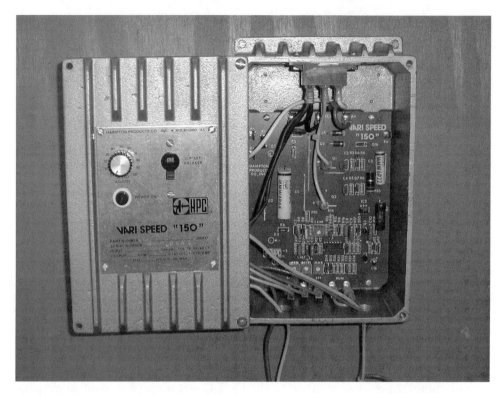

Figure 3-1: *An HPC Vari Speed 150 SCR armature voltage controller.*

feedback control. Because a detailed explanation of this method of control was given in Chapter 2, we will not repeat the theory of the feedback circuit here. We will begin instead by focusing on the lower portion of Figure 3-2, where the circuit that consists of the null detector, pulse shaper, and sawtooth generator are located.

What Do the Null Detector, Pulse Shaper, and Sawtooth Generator Do?

Begin by looking at the null detector in Figure 3-3 on page 26. Note that the AC voltage is rectified by the full-wave bridge circuit consisting of CR7–CR10. This circuit causes an unfiltered, pulsating DC voltage to appear across Zener diode Z1, R28, and the LED of optocoupler U10.

First we will consider what happens as this pulsating DC voltage rises from 0V to its peak value. The LED of U10 will not conduct until the DC voltage reaches the turn-on voltage of the LED. Assume that this voltage is approximately 1.5V. As the DC voltage increases from 0V to 1.5V, U10 is turned off.

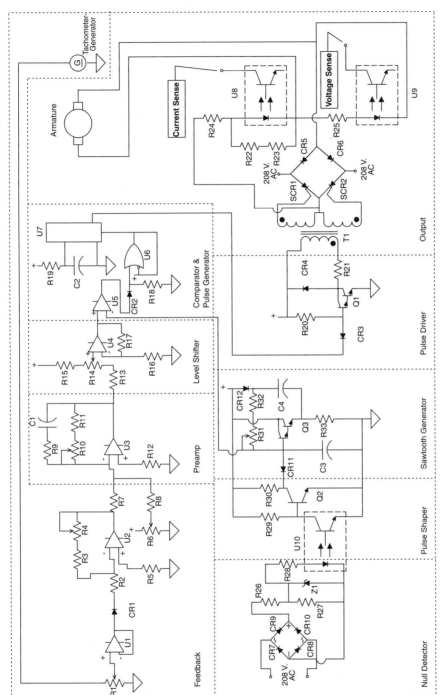

Figure 3-2: *Schematic of an SCR armature voltage controller.*

25

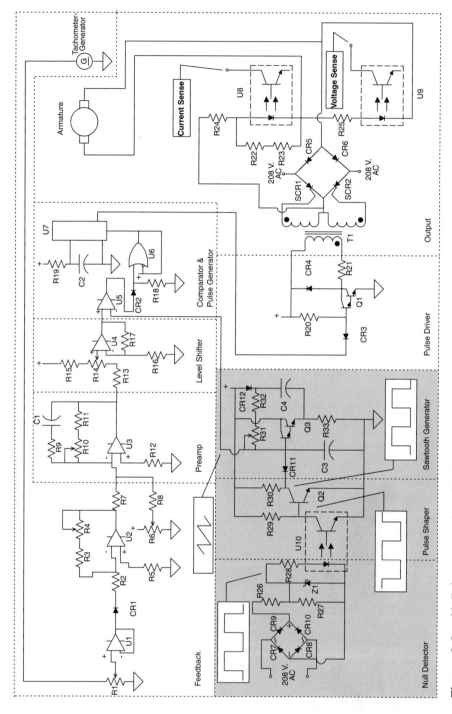

Figure 3-3: Null detector, pulse shaper, and sawtooth generator sections.

26

Therefore the output of U10 will be a positive pulse. This positive pulse at the output of U10 causes Q2 to turn on. When Q2 conducts, a negative pulse will appear at the collector of Q2. This negative pulse resets the output of the sawtooth generator Q3 back to 0.

Now we will look at what happens when the LED of U10 turns on. As the pulsating DC voltage rises, eventually it reaches a value of 1.5V. The voltage across the LED of U10 will never exceed the voltage rating of Zener diode Z1. At this point, the LED of U10 turns on and conducts, causing the output of U10 to supply a negative pulse that turns Q2 off. When Q2 is turned off, the sawtooth generator outputs a sawtooth waveform to the inverting input of U5 (the comparator). The sawtooth generator will continue to output this waveform until it is reset when Q2 turns on, that is, when the pulsating DC voltage falls below 1.5V and continues to decrease to 0V. In this way the pulsating DC voltage causes the sawtooth generator to provide the comparator with a reference signal that is used to vary the firing angle of the SCRs that control the armature voltage.

Next, look at U5, the comparator in Figure 3-4 on the next page. The comparator has two inputs. One, which comes from the feedback circuit, is a measure of the speed of the armature. The other, as we just learned, is a ramp signal from the sawtooth generator. The noninverting input from the feedback amplifier will be constant if the load on the armature is constant. The inverting input is a rising amplitude signal from the sawtooth generator. Initially, the noninverting input causes the output of U5 to be positive. However, as the inverting input ramp signal climbs, at a given point the amplitudes of both inputs become equal. At this point the output of U5 switches off and becomes a rectangular pulse. The width of this pulse is determined by the length of time it takes for both inputs to become equal. This is what determines whether the SCR conducts earlier or later within the positive half of the cycle. This process is similar to that of the comparator stage discussed in Chapter 2.

What Is the Function of the Pulse Generator?

The **pulse generator,** shown in Figure 3-5 (see page 29), is an integrated circuit that contains a "one-shot" or monostable multivibrator. The purpose of the pulse generator is to trigger on the output of the comparator and thus provide a narrow pulse to Q1, the pulse driver.

What happens next? Referring to Figure 3-6 (see page 30), recall that Q1 will not conduct until a positive pulse appears at its base. When this occurs, current will flow through the primary of T1. Q1 will only conduct for the duration of the

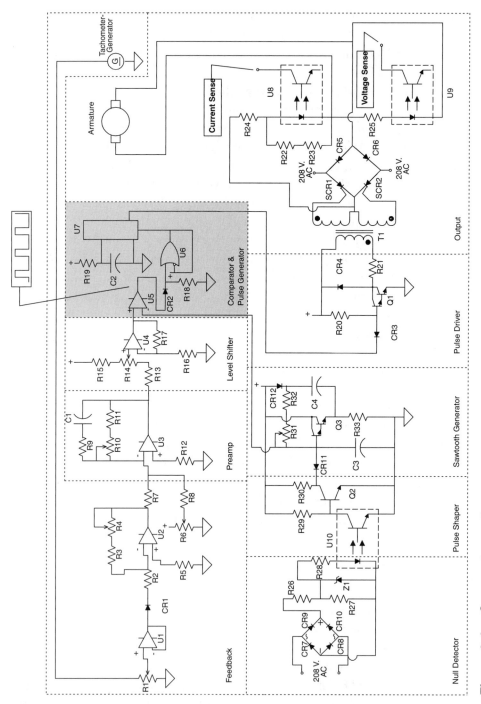

Figure 3-4: *Comparator section.*

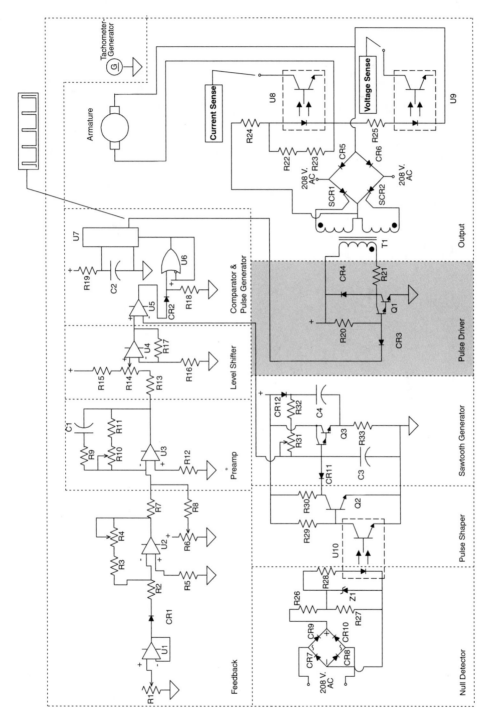

Figure 3-5: *Pulse-generator section.*

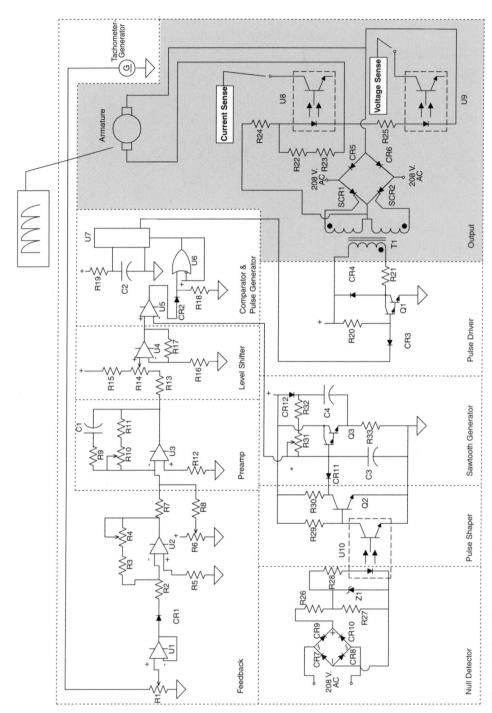

Figure 3-6: Output section.

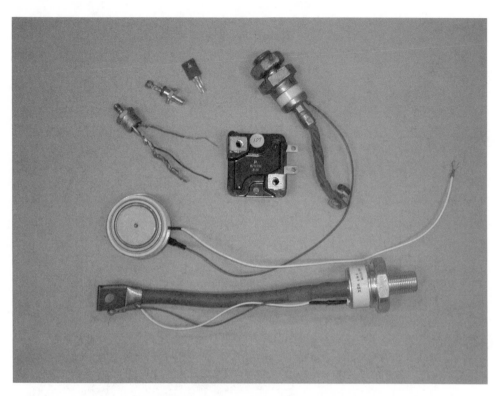

Figure 3-7: *Various SCR packages.*

pulse applied to its base. The current flowing through the primary of T1 will cause a current flow in the secondary of T1. Therefore the gates of SCR1 and SCR2 receive a trigger pulse simultaneously. Figure 3-7 shows some of the common types of SCRs in use. However, note that the anodes of SCR1 and SCR2 are connected to an AC source. As a result, when the anode of SCR1 is positive, the anode of SCR2 is negative and vice versa. Thus when SCR1 and SCR2 are triggered, only the SCR with the positive anode conducts, so the SCRs conduct alternately. Consequently, DC current flows through the armature of the DC motor. Figure 3-8 shows the circuitry used to control the SCRs while Figure 3-9 (see page 33) shows the circuitry used to rectify the AC with SCRs.

Now let's bring this all together. Figure 3-10 shows the waveforms at various points in the comparator and Pulse Generator, Pulse Driver, and Output stages. These waveforms have been lined up vertically so that you understand the timing of the waveforms that must occur in order for the output SCRs to trigger at the appropriate time. Here's how it works. The waveform in Figure 3-10a shows

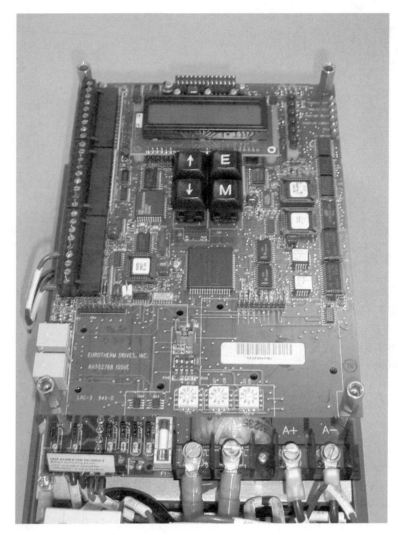

Figure 3-8:   *Control board assembly.*

the sawtooth waveform at the inverting input of U5, the comparator. At the
same time, Figure 3-10a also shows the DC voltage that is applied to the noninverting input of comparator U5. Recall that this voltage is the feedback voltage from the tachometer-generator, and is an indication of the armature speed.
The DC voltage level shown is representative of a motor speed of 1800 RPM.
These two inputs cause the output of the comparator to produce a pulsating DC
as shown in Figure 3-10b. Notice that the pulse is initially positive, and is
switched off when the two inputs of the comparator become equal. The output

Figure 3-9: *SCR power rectifier section.*

of U5 remains off until the sawtooth generator resets. At this point, the output of U5 switches on, and remains on until the sawtooth signal is equal to the feed-back voltage.

The pulsating output of U5 is fed through **diode** CR2 to the input of U6. Because of CR2, the input signal to U6 will be inverted compared to the output of U5. This is shown in Figure 3-10c. The positive pulses from the output of U6 trigger the one-shot, U7. The output of the one-shot, U7, will appear in phase with the input pulses as seen in Figure 3-10d.

The output of the one-shot, U7, is fed through diode CR3 to the base of the Darlington transistor, Q1. Notice in Figure 3-10e that the collector of Q1 is in phase with the output of the one-shot. When Q1 conducts, current will flow through the pulse transformer, T1.

When current flows through pulse transformer T1, a pulse appears at the gates of SCR1 and SCR2. This is shown in Figure 3-10f. Recall that only the SCR that is properly biased will conduct. The SCR that is triggered and prop-erly biased will conduct until the AC voltage drops to the zero crossing point.

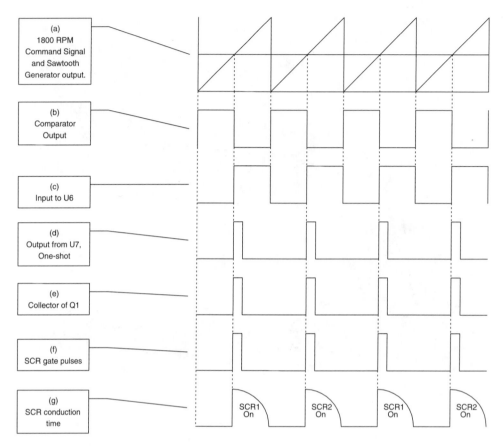

Figure 3-10: *Waveform timing and SCR conduction time at 1800 RPM.*

The next pulse from T1 will cause the other SCR to conduct. Therefore, the SCRs will conduct alternately as shown in Figure 3-10g.

How Does This Process Automatically Vary DC Motor Speed?

Let's assume that an increase in load has caused the motor speed to decrease to 1000 RPM. Refer to Figure 3-11. The waveform in Figure 3-11a shows the saw-tooth waveform at the inverting input of U5, the comparator. At the same time, Figure 3-11a also shows the DC voltage that is applied to the non-inverting input of comparator U5. The DC voltage level shown is representative of a motor speed of 1000 RPM. Notice that this level is lower than that shown in Figure 3-10a. This is a result of the slower motor speed and lower output voltage

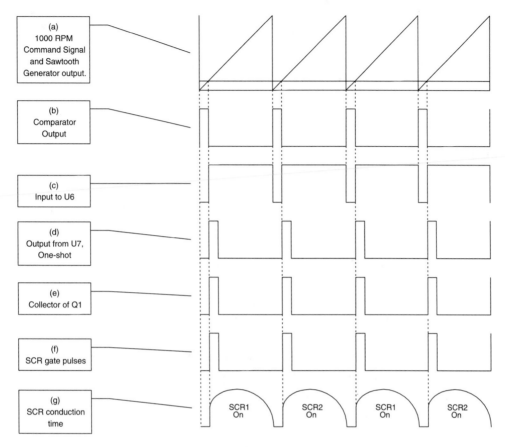

Figure 3-11: *Waveform timing and SCR conduction time at 1000 RPM.*

from the tachometer-generator. These two inputs cause the output of the comparator to produce a pulsating DC as shown in Figure 3-11b. Notice that the pulse is initially positive, and is switched off when the two inputs of the comparator become equal. The output of U5 remains off until the sawtooth generator resets. At this point, the output of U5 switches on, and remains on until the sawtooth signal is equal to the feedback voltage. Compare the width of the positive pulses in Figure 3-11b with those in Figure 3-10b. Notice that the positive pulses are narrower in Figure 3-11b.

The pulsating output of U5 is fed through diode CR2 to the input of U6. Because of CR2, the input signal to U6 will be inverted compared to the output of U5. This is shown in Figure 3-11c. Again, compare the positive pulses in Figure 3-11c with those shown in Figure 3-10c. Notice that the positive pulses are wider in Figure 3-11c than those shown in Figure 3-10c. The positive pulses from the output of U6 trigger the one-shot, U7. The output of the one-shot, U7, will appear in phase with the input pulses as seen in Figure 3-11d. Notice that the output pulse of the one-shot now occurs earlier than it did in Figure 3-10d. This will cause the SCRs to fire sooner, causing them to conduct for a longer period of time. Let's see if this is what occurs.

The output of the one-shot, U7, is fed through diode CR3 to the base of the Darlington transistor, Q1. Notice in Figure 3-11e that the collector of Q1 is in phase with the output of the one-shot. When Q1 conducts, current will flow through the pulse transformer, T1.

When current flows through pulse transformer T1, a pulse appears at the gates of SCR1 and SCR2. This is shown in Figure 3-11f. Recall that only the SCR that is properly biased will conduct. The SCR that is triggered and properly biased will conduct until the AC voltage drops to the zero crossing point. The next pulse from T1 will cause the other SCR to conduct. Therefore, the SCRs will conduct alternately as shown in Figure 3-11g. Notice that the SCRs conduct for a longer amount of time as compared to the conduction of the SCRs in Figure 3-10g. This results in a higher average voltage applied to the armature of the DC motor. This will cause the speed of the DC motor to increase.

Let's assume that a decrease in load has caused the motor speed to increase to 2600 RPM. Refer to Figure 3-12. The waveform in Figure 3-12a shows the sawtooth waveform at the inverting input of U5, the comparator. At the same time, Figure 3-12a also shows the DC voltage that is applied to the non-inverting input of comparator U5. The DC voltage level shown is representative of a motor speed of 2600 RPM. Notice that this level is higher than that shown in Figure 3-10a. This is a result of the faster motor speed and higher output voltage from the tachometer-generator. These two inputs cause the output of the comparator to produce a pulsating DC as shown in Figure 3-12b. Notice that the pulse is initially positive, and is switched off when the two inputs of the comparator become equal. The output of U5 remains off until the sawtooth generator resets. At this point, the output of U5 switches on, and remains on until the sawtooth signal is equal to the feedback voltage. Compare the width of the positive pulses in Figure 3-12b with those in Figure 3-10b. Notice that the positive pulses are wider in Figure 3-12b.

The pulsating output of U5 is fed through diode CR2 to the input of U6. Because of CR2, the input signal to U6 will be inverted compared to the output of U5. This is shown in Figure 3-12c. Again, compare the positive pulses in Figure 3-12c with those shown in Figure 3-10c. Notice that the positive pulses are narrower in Figure 3-12c than those shown in Figure 3-10c. The positive pulses from the output of U6 trigger the one-shot, U7. The output of the one-shot, U7,

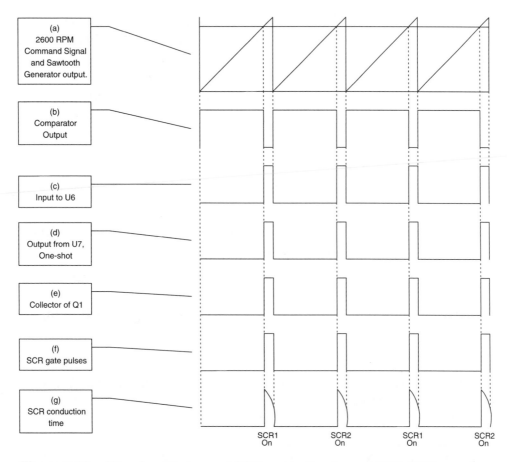

Figure 3-12: *Waveform timing and SCR conduction time at 2600 RPM.*

will appear in phase with the input pulses as seen in Figure 3-12d. Notice that the output pulse of the one-shot now occurs later than it did in Figure 3-10d. This will cause the SCRs to fire later, causing them to conduct for a shorter period of time. Let's see if this is what occurs.

The output of the one-shot, U7, is fed through diode CR3 to the base of the Darlington transistor, Q1. Notice in Figure 3-12e that the collector of Q1 is in phase with the output of the one-shot. When Q1 conducts, current will flow through the pulse transformer, T1.

When current flows through pulse transformer T1, a pulse appears at the gates of SCR1 and SCR2. This is shown in Figure 3-12f. Recall that only the SCR that is properly biased will conduct. The SCR that is triggered and properly biased will conduct until the AC voltage drops to the zero crossing point.

The next pulse from T1 will cause the other SCR to conduct. Therefore, the SCRs will conduct alternately as shown in Figure 3-12g. Notice that the SCRs conduct for a shorter amount of time as compared to the conduction of the SCRs in Figure 3-10g. This results in a lower average voltage applied to the armature of the DC motor. This will cause the speed of the DC motor to decrease.

Review Questions

1. Explain the effect of increasing the armature voltage on the speed of a DC motor.

2. List some of the differences between the SCR armature voltage controller and the switching amplifier field current controller.

3. List some of the similarities between the SCR armature voltage controller and the switching amplifier field current controller.

4. Explain the process by which the SCR armature voltage controller adjusts the speed of a DC motor?

5. Explain how the firing point of the SCRs is adjusted.

6. **True or False?** The SCR armature voltage controller is the most commonly used method for controlling the speed of a DC motor.

Challenge Yourself

Refer to Figures D-1 and D-2 in Appendix D as you answer these questions.

1. Is this type of DC drive a switching amplifier field current controller or an SCR armature voltage controller?

2. In Figure D-2 a jumper labeled "ARM" is located below and to the left of R16 on the right side of the schematic. What is the purpose of this jumper?

3. What occurs at pin #6 of op-amp A2-B?

4. In Figure D-1, what is the purpose of op-amp A1-B?

5. In Figure D-1, what is the purpose of transformer T2?

Chapter 4

Brushless DC Motor Controllers

OBJECTIVES

After completing this chapter, you will be able to:

- Discuss the theory of operation of a brushless DC motor.
- Discuss the differences between a conventional DC motor and a brushless DC motor.
- List some advantages of the brushless DC motor.
- Discuss two different methods of sensing rotor position.

In this chapter we will discuss the brushless DC motor and controller. We will also consider the three-phase and four-phase methods of control and learn about magnetic and optical position sensing.

What Is a Brushless DC Motor, and How Does It Operate?

As its name implies, a **brushless DC motor (BLDC motor)** is a DC motor that does not contain **brushes.** To understand its operation, let's review how a "regular" DC motor operates.

Remember that a DC motor has a **commutator.** (See Figure 4-1.) Electrical connections are made to the commutator by means of one or more sets of brushes. The commutator reverses the polarity of the DC voltage applied to the rotor, thus causing the rotor to turn in one direction as a result of the attraction of the magnetic fields in the stator and rotor.

Figure 4-2A (at the top of page 41) shows how the stator is connected to a DC source. Notice the polarity of the stator windings. The connections to the stator do not change; the stator always has the same magnetic polarity. Now watch what happens as the rotor revolves. When the DC is applied to the rotor shown

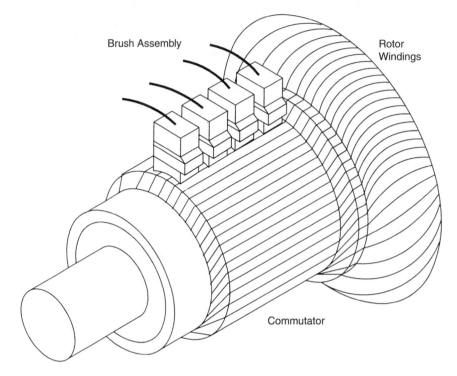

Brush Assembly

Rotor
Windings

Commutator

Figure 4-1:   *Commutator of a DC motor.*

in this figure, the rotor's magnetic field causes it to be attracted to the stator winding with the opposite magnetic polarity, and the rotor consequently turns in that direction.

Next look at Figures 4-2B through 4-2D (see pages 41–42). The commutator reverses the polarity of the DC voltage applied to the rotor, causing the rotor to continue rotating in the same direction. The strength of the magnetic field in the rotor controls the rotor's speed, usually via a variable resistor in the rotor circuit. To change the direction of the rotor rotation, we can reverse the DC connections either to the rotor or to the stator.

The brushless DC motor addresses two major shortcomings of the regular DC motor. The first shortcoming is the maintenance cost associated with brush replacement and commutator wear in regular DC motors. The second is the arcing that the brushes produce on the commutator, causing not only electrical "noise" but also problems when such a motor is used in an explosive environment.

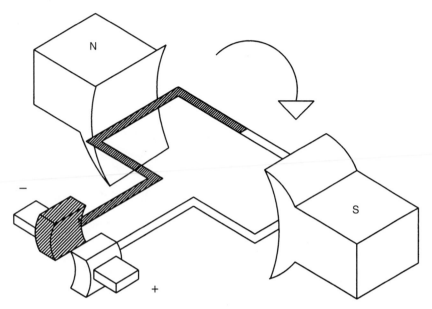

Figure 4-2A: *Commutator at 0°.*

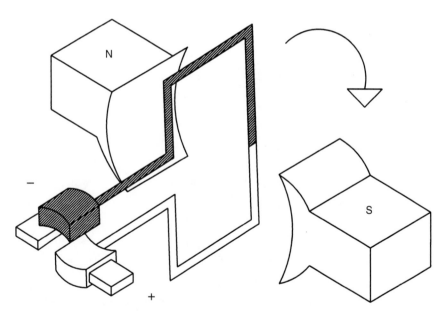

Figure 4-2B: *Commutator at 90°.*

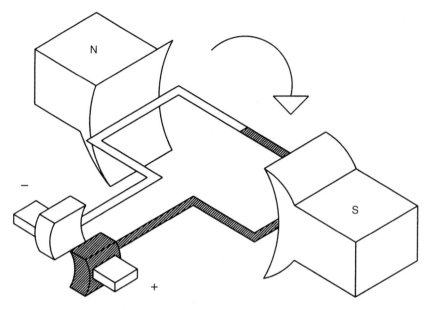

Figure 4-2C: Commutator at 180°.

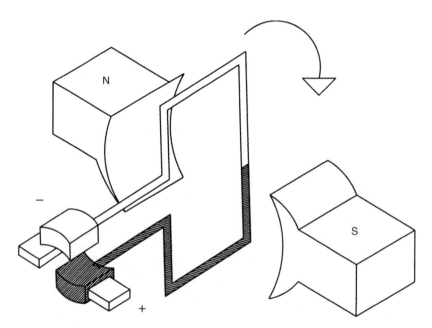

Figure 4-2D: Commutator at 270°.

42

How Is the BLDC Motor Constructed?

The BLDC motor is a permanent magnet motor with electronic commutation. Let's look at this in more detail. As shown in Figure 4-3, the rotor has a permanent magnet instead of rotor windings. This means no electrical connections to the rotor and therefore no need for a commutator. Immediately we have eliminated many of the maintenance requirements and shortcomings of the regular DC motor. The stator of the BLDC motor has multiple windings; the figure shows a simplified version. Among the electronic devices used to provide commutation are the **magnetic encoder,** known as the **Hall-effect device,** or an **optical encoder,** which consists of a **phototransistor** and a shutter attached to the rotor shaft.

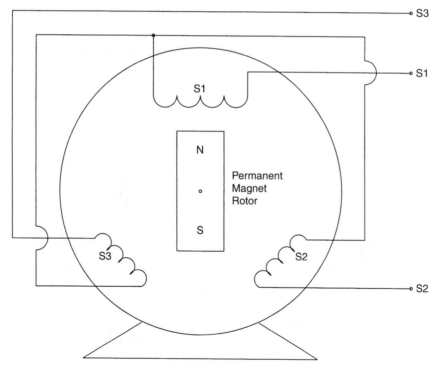

Figure 4-3: Brushless DC motor construction.

How Does the Three-Phase (Optical Sensing) BLDC Motor Controller Operate?

Three phototransistors are positioned around the shaft of the BLDC motor, spaced 120° apart. The shutter, attached to the rotor shaft, blocks light from all but one of the phototransistors at a time. As the rotor revolves, the shutter rotates to allow light to strike a different phototransistor. In Figure 4-4, the shutter allows light to strike phototransistor Q1. The light turns on Q1, which in turn causes Q4 to conduct, allowing current to flow through stator winding S1. The rotor then rotates clockwise and aligns itself with stator winding S1. As the rotor rotates, so does the shutter. This rotation causes phototransistor Q1 to be blocked and phototransistor Q2 to be exposed to light. When light strikes phototransistor Q2, current is allowed to flow to and turn on Q5. This event causes current to flow through stator winding S2. The rotor continues turning clockwise until it has aligned itself with stator winding S2. As the shutter rotates simultaneously to block phototransistor Q2, it exposes phototransistor Q3. Stator winding S3 now receives current because Q3 turns on Q6. The rotor continues to rotate continuously clockwise. Naturally, this occurs very rapidly, and the rotor turns smoothly. This arrangement is known as optically sensing rotor position.

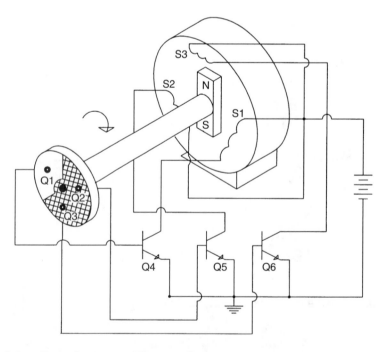

Figure 4-4: *Optical rotor position sensing.*

How Does the Four-Phase (Magnetic Sensing) BLDC Motor Controller Operate?

As we just saw, the three-phase BLDC motor controller senses the rotor position optically. Now we will consider how the four-phase BLDC motor senses the rotor position magnetically, using a different type of sensor, a Hall-effect sensor. As Figure 4-5 shows, we have added a fourth stator winding and two Hall-effect sensors. This device works in the following way. With the rotor in the position shown in the figure, Hall-effect sensor H1 produces an output voltage. This voltage turns on Q1. When Q1 conducts, current flows through stator winding S1, causing the rotor to rotate clockwise. As the rotor rotates away from Hall-effect sensor H1, Q1 is turned off. At the same time, the rotor is aligning itself with Hall-effect sensor H2. This causes H2 to produce a voltage that turns on transistor Q2. When Q2 conducts, current flows through stator winding S2, causing the rotor to continue rotating clockwise and move away from sensor H2. Now the opposite polarity of the permanent magnet rotor appears under sensor H1. This causes H1 to output a voltage to transistor Q3. With Q3 thus turned on, current flows through stator winding S3. Again, the rotor continues to rotate

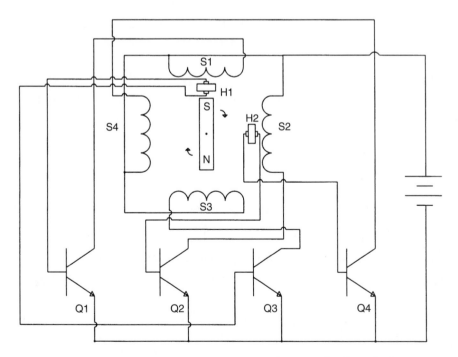

Figure 4-5: *Magnetic rotor position sensing using Hall-effect devices.*

clockwise. After moving away from H1, the rotor aligns, with its opposite polarity (in relation to the original conditions), at H2. The resulting voltage from H2 turns on Q4, and current flows through stator winding S4. The rotor continues to move clockwise, having completed one revolution. Again, this process occurs very rapidly, and the rotor revolves smoothly and continuously.

How Do We Control the Direction of Rotation?

The BLDC motor controller has a logic circuit that determines which transistors (controlled either optically or magnetically) can conduct, and in what order. If, in our last example, we had allowed transistor Q4 to conduct first, followed by Q3, Q2, and finally Q1, the rotor would have revolved counterclockwise. Therefore the logic circuitry of the controller for the BLDC motor determines the direction of rotation. As an additional benefit, this same logic circuitry also controls the power supply to the BLDC motor.

How Do We Control the Speed of the BLDC Motor?

The logic circuitry mentioned in the previous section also controls the speed of the BLDC motor by determining how often and for what length of time the transistors are turned on. Because the logic circuitry controls how often the transistors are turned on, we can say that it determines their frequency of operation. This function is known as **pulse frequency control.** The logic circuitry's control over the length of time that the transistors conduct is called **pulse width control.**

Review Questions

1. What do the letters BLDCl motor stand for?

2. Explain the purpose of a commutator in a conventional DC motor.

3. Describe how commutation is accomplished in a BLDC motor.

4. **True or False?** Two disadvantages of the conventional DC motor are cost and weight.

5. List two advantages of the BLDC motor and explain why they are advantages.

6. Describe the operation of a Hall-effect sensor.

7. A phototransistor

   a. is used only in photographic equipment.

   b. conducts in the presence of light.

   c. emits light when properly biased.

   d. continues to conduct even after the light source has been removed.

8. Describe the purpose of the shutter in a BLDC motor.

9. Explain the operation of a BLDC motor controller.

10. Describe what must be done to vary the direction of rotation of a BLDC motor.

Chapter 5

Braking

OBJECTIVES

After completing this chapter, you will be able to:

- List six different types of braking.
- Discuss the differences among the six different types of braking.
- Discuss the advantages and disadvantages of the six different types of braking.

In this chapter we discuss several methods of decelerating a motor, or braking. For a DC drive or even an inverter drive to operate properly and provide precise speed control, it must offer some form of braking. In this chapter we look at six different methods of braking: (1) mechanical braking, (2) ramping, (3) plugging, (4) DC injection, (5) dynamic braking, and (6) regeneration.

What Is a Mechanical Brake, and How Does It Work?

Mechanical brakes, also called magnetic or friction brakes, work in much the same way as the brakes in your car or truck. (See Figure 5-1, page 50)

Basically, a mechanical brake uses two brake shoes, with a friction coating or pad, mounted in close proximity to a drum attached to the motor's shaft. Connected to the shoes is a tension spring that keeps the shoes in contact with the drum. This contact provides braking action.

If the motor is to turn, we must lift the shoes from the drum. A solenoid attached to a lever mechanism accomplishes this task. In order for the motor to operate, the solenoid is energized, which moves the lever, lifting the brake shoes and allowing the motor to revolve. When the motor stops the solenoid is deenergized, and the spring's tension forces the brake shoes into contact with the drum. The resulting friction rapidly brings the motor to a stop.

Figure 5-1: *An electro-mechanical friction brake.*

What Is Ramping?

Electronic variable speed drives can gradually decrease the current, voltage, and frequency supplied to a motor, reducing these parameters automatically and in the correct proportions. The user can program or otherwise adjust how rapidly this reduction occurs. This procedure, called **ramping,** gradually slows the motor in a controlled fashion. Note, however, that this form of braking is not very effective for high inertia loads, which may cause **overhauling,** that is, continued rotation of the motor even after it has been disconnected, due to centrifugal force.

What Is Plugging?

Plugging should be reserved for emergency stop situations only. This method of braking exerts a very strenuous force on the motor and can cause severe damage to the motor if this method is used frequently.

Plugging reverses the direction of a motor while it is still in motion and is thus akin to throwing your car's transmission into reverse while it is traveling forward. You can imagine the results! Again, this form of braking should be reserved for emergency stop applications only!

What Is DC Injection?

DC injection applies only to AC induction motors. Remember that in an AC induction motor it is essentially a rotating magnetic field that causes the rotor to revolve. Applying a DC current to the phase windings of this motor causes a fixed magnetic field to form. This field does not revolve and, in fact, causes the field to behave as a permanent electromagnet. The rotor, naturally attracted to this fixed magnetic field, locks in place. This method of braking is also quite hard on a motor, and therefore its use should be limited.

What Is Dynamic Braking?

Remember that as the armature of a motor revolves within a magnetic field, it produces a back EMF, or CEMF. This counter EMF opposes the applied voltage and thus helps to decrease the armature current. If power is removed from the motor, you would expect the motor to coast to a stop.

However, connecting large resistors across the motor produces even better braking performance. Be very careful when working around these resistors. As the braking action occurs, these resistors must dissipate large amounts of energy. This will cause the resistors to become very hot and may produce burns upon contact. Now, as power is reduced, the motor spins down. At the same time, the motor's armature is still revolving inside a magnetic field. This rotation causes the motor to act as a generator, producing the CEMF. Because this generated CEMF opposes the voltage applied to the motor, the motor produces a counter torque that opposes the original direction of rotation and thus applies further braking action, called **dynamic braking,** to the motor. The downside of this effect is that, as the braking action occurs and the motor speed slows, less CEMF is produced. Thus the braking action becomes weaker as the motor's speed decreases. For this reason auxiliary mechanical braking is often necessary in conjunction with dynamic braking to bring the load to a complete stop.

What Is Regenerative Braking?

Regenerative braking is very similar to dynamic braking. The major difference is that with regenerative braking the motor is not fully disconnected from the power source. As the motor produces its CEMF, the resulting energy is fed back into the power supply. This effect can lower operating expenses because it produces basically "free" energy. Regenerative braking is the preferred method of braking used with electronic variable speed drives. Although regenerative braking is more expensive than other braking alternatives, it does produce some useable energy that can help offset the additional associated expense.

Review Questions

1. What, if any, differences exist between plugging and DC injection?

2. Explain why plugging should be used only for emergency stop applications.

3. The type of braking that uses large resistors to dissipate energy from the load is

   a. plugging.

   b. dynamic braking.

   c. mechanical braking.

   d. regenerative braking.

4. Which type of braking is most commonly used with electronic variable speed drives? Justify your answer.

5. **True or False?** You must use additional braking, such as a mechanical brake, to brake a motor completely when using dynamic braking.

6. Describe how regenerative braking and dynamic braking differ.

7. Can you use DC injection braking repeatedly? Why or why not?

8. Which type of braking feeds energy back into the power supply? Explain how this is accomplished.

9. Describe how a mechanical brake works.

10. Describe ramping.

Chapter 6

Chopper Circuits

OBJECTIVES

After completing this chapter, you will be able to:

- Describe the operation of a chopper circuit.

- Discuss the differences between a buck chopper and a boost chopper.

- List the four quadrants of motor operation.

- Explain the differences between the four quadrants of motor operation.

- Describe the function of a free-wheeling diode.

In this chapter we introduce yet another circuit that can transform steady DC into pulsating DC: the chopper. We will focus on two of the various types of chopper circuits: the buck chopper and the boost chopper.

What Is a Chopper?

A chopper is a circuit that uses very fast electronic switches. Sometimes these switches consist of transistors. At other times SCRs are used. But in this chapter we will study the use of metal oxide semiconductor field effect transistors (MOSFETs). **MOSFETs** switch the applied DC on and off very rapidly. By using them we can create a DC output level that will be either lower or higher than the applied DC. If the chopper's output is *lower* than the applied DC, the chopper is called a **buck chopper,** or **step-down chopper.** If the chopper output is *higher* than the applied DC, the chopper is called a **boost chopper,** or **step-up chopper.**

How Does a Buck Chopper Work?

In Figure 6-1, notice that the gate of MOSFET Q1 is connected to a control signal source. The control signal emitted by this source is a pulse-width modulated (PWM) signal. We will begin by setting the duty cycle of the PWM signal to 50%. (Remember that the duty cycle is the ratio of the time of the "on" pulse to the time of one cycle. Therefore a pulse set at a 50% duty cycle is on for the duration of half one cycle.)

When the control signal pulse is on it is applied to the gate of the MOSFET, and Q1 conducts as a result. However, Q1 will only conduct for half of one cycle. While Q1 conducts, the armature of the DC motor is connected to the DC supply. This connection causes diode CR1, a **free-wheeling diode,** to appear as an open circuit because it is reverse biased. Therefore the current flowing through the armature of the DC motor increases. During the next half-cycle, this pulse to the gate of Q1 is switched off, so Q1 is also turned off. As a result, the armature of the DC motor is no longer connected to the DC supply, and diode CR1 provides a discharge path for the collapsing magnetic field of the armature.

The output voltage of a buck chopper is proportional to the duty cycle of the control signal. If we increase the duty cycle (thus making the "on" pulse wider), the average value of the chopper's output voltage also increases. Likewise, if we decrease the duty cycle (making the "on" pulse narrower), the average value of the chopper's output voltage decreases as well. Varying the frequency of the control signal has no effect on the level of output voltage and current. However, a higher frequency control signal produces less ripple frequency in the output voltage and current, and thus results in smoother motor operation. Because of losses in the MOSFET, at no time will the output voltage of the buck chopper exceed the DC supply voltage. That is the reason why this particular chopper is known as a buck or step-down chopper.

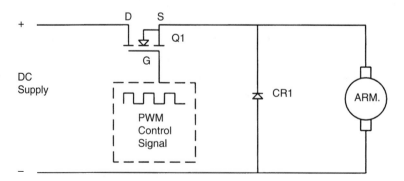

Figure 6-1: *Buck or step-down chopper.*

What Is a Boost Chopper?

The boost chopper shown in Figure 6-2 again uses a PWM signal to turn the gate of MOSFET Q1 on and off. Let's begin with a 50% duty cycle and a charge on capacitor C1. When the signal pulse is on, it is applied to the gate of MOSFET Q1, and Q1 is also turned on. With Q1 turned on, a very low voltage will develop across the source to drain leads of Q1, which in turn allows current to flow through inductor L1 and thus builds up a magnetic field. Because diode CR1 is now reverse biased, no current will flow through it. Therefore capacitor C1 discharges through the armature of the DC motor, causing the armature current, voltage, and motor speed to decrease as the charge on C1 decreases.

When the signal pulse to Q1 is switched off during the next half-cycle, Q1 is also turned off, causing the source-to-drain voltage of Q1 to increase. As a result, CR1 now becomes forward biased, allowing the magnetic field of inductor L1 to collapse. The discharge current from L1, along with the supply current, recharges C1 and also increases the amount of current flowing through the armature of the DC motor. In turn, this increased armature current causes corresponding increases in armature voltage and motor speed.

To sum up, if Q1 is turned on for a *longer* period of time, motor speed decreases. If, on the other hand, Q1 is turned on for a *shorter* period of time, motor speed increases. Again, varying the frequency of the control signal has no effect on the level of output voltage and current. However, a *higher* frequency control signal produces less ripple frequency in the output voltage and current, resulting in smoother motor operation. Because the inductor voltage boosts, or supplements, the supply voltage, the output voltage can be higher than the input voltage. That is the reason why this circuit is known as a boost or step-up chopper.

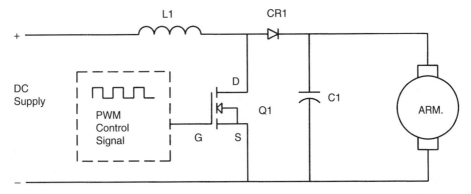

Figure 6-2: *Boost or step-up chopper.*

What Are the Limitations of These Drives?

We must realize that current will only flow in one direction, from source to load, in both of these drives. Thus both of these controls have some shortcomings. Most notable is the fact that speed control is difficult with high-inertia loads, which may cause overhauling.

Overhauling occurs when the load has enough momentum to cause the motor rotor to continue rotating even after the motor has been disconnected. A flywheel is an example of this type of load. The type of drive in which this phenomenon occurs is a **one-quadrant drive.**

What Is a One-Quadrant Drive?

First, look at the four quadrants shown in Figure 6-3. Note the polarities of the current and voltage in each quadrant. In quadrant 1 the current and the voltage are both positive. When a motor is operating in quadrant 1, we say that the motor is *motoring,* that is, driving a load and turning in the forward direction.

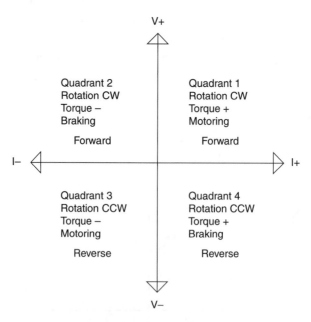

Figure 6-3: The four quadrants of motor operation.

What Is a Two-Quadrant Drive?

Look at quadrant 3 in Figure 6-4. Note that in this quadrant the current and the voltage are both negative. A motor operating in quadrant 3 is still motoring, but it is now operating in reverse.

Do Motors Ever Operate in More Than One Quadrant?

Sometimes a motor must operate in both quadrants 1 and 3. This is necessary when a motor must drive a load both forward and in reverse. Examples are the motor in an electric vehicle that runs forward and backward or in a crane that must raise and lower a load.

What Happens in Quadrants 2 and 4?

In Figure 6-5, look first at quadrant 2. Note that although the voltage is still positive, the current is negative. We say that a motor operating in quadrant 2 is *regenerating*. During this time the motor provides energy back to the power

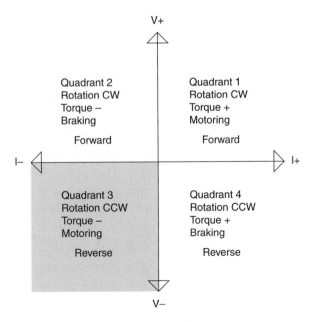

Figure 6-4:   *Quadrant 3: Motoring in reverse.*

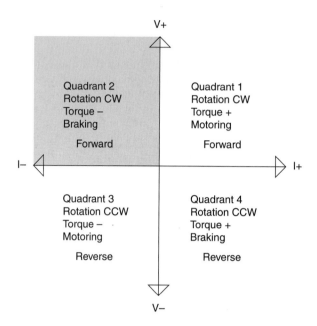

Figure 6-5: *Quadrant 2: Braking in forward.*

source while rotating forward. In other words, the motor acts like a generator. This regenerated energy provides braking for the motor, affording us better control over the motor's speed. However, operating in quadrant 2 does not allow for braking when the motor turns in reverse.

Now look at quadrant 4 in Figure 6-6. Here the voltage is negative, and the current is positive. Motor operation in quadrant 4 produces regenerative braking action when the motor turns in reverse. Think about the electric vehicle example. We want the vehicle to be able not only to travel forward and backward, but also to stop in either direction of travel. Therefore a DC motor for an electric vehicle must be able to operate in all four quadrants. Likewise, in the example of a crane, not only must we be able to lift and lower a load, we must also be able to control the speed at which it raises and lowers the load. Again, the DC motor must therefore operate in all four quadrants. In Chapter 7 we will learn about a drive that operates in all four quadrants: a four-quadrant chopper.

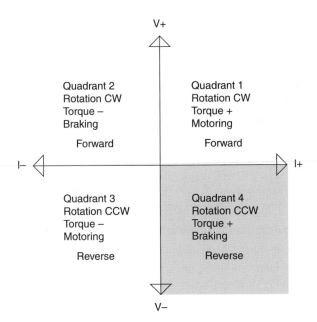

Figure 6-6:   *Quadrant 4: Braking in reverse.*

Review Questions

1. Explain what a chopper does.

2. What is the purpose of the free-wheeling diode in a buck chopper?

3. The component responsible for producing the higher output levels in a boost chopper is

    a. a capacitor.

    b. a MOSFET.

    c. an inductor.

    d. a and c.

    e. a, b, and c.

4. **True or False?** Pulse width modulation is used with the MOSFETs in a chopper.

5. Is the length of time that a MOSFET is switched on proportional or inversely proportional to the output voltage level of a buck chopper? Explain your answer.

6. Is the length of time that a MOSFET is switched on proportional or inversely proportional to the output voltage level of a boost chopper? Explain your answer.

7. **True or False?** Varying the frequency of the control signal does not affect the output voltage level of a buck chopper.

8. **True or False?** Varying the frequency of the control signal does affect the output voltage level of a boost chopper.

9. In which operating quadrant does motoring in reverse occur?

   a. First.

   b. Second.

   c. Third.

   d. Fourth.

10. In which quadrant(s) must a motor operate to achieve both forward and reverse operation with braking capabilities in both directions?

   a. First.

   b. Second.

   c. Third.

   d. Fourth.

Chapter 7

The Four-Quadrant Chopper

OBJECTIVES

After completing this chapter, you will be able to:

- Discuss the operation of a four-quadrant chopper.
- Discuss the differences between a four-quadrant chopper and a one- or two-quadrant chopper.

In this chapter we will see how a four-quadrant chopper operates and learn how it provides control in all four operating quadrants.

What Is a Four-Quadrant Chopper?

As its name implies, a **four-quadrant chopper** operates in all four quadrants. That is, the motor operates both forward and in reverse, and it also has regenerative braking capabilities in both directions.

What Is Regenerative Braking?

Regenerative braking occurs when the motor still rotates even after power has been removed. When power is removed from a motor, the motor continues to rotate until friction and windage slow it to a stop. While the motor is rotating, the magnetic fields in the armature and stator are collapsing. The motion of the armature through these collapsing magnetic fields generates a voltage and a current. The polarity of this generated voltage and current is the opposite of the polarity of the voltage and current that were originally applied to the DC motor. This opposite polarity tries to make the motor turn in the opposite direction, in effect braking the motor. Because the braking force is a result of the generated voltage and current, it is called regenerative braking.

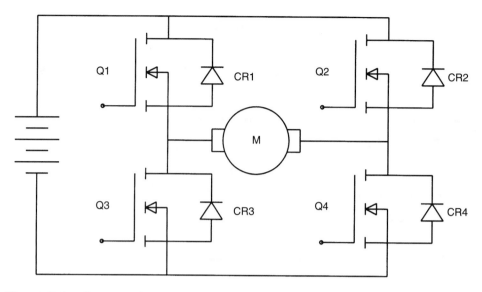

Figure 7-1: *Four-quadrant chopper.*

How Does a Four-Quadrant Chopper Operate?

In Figure 7-1, notice the four MOSFETs: Q1, Q2, Q3, and Q4. Also notice that each MOSFET has an associated free-wheeling diode: CR1, CR2, CR3, and CR4. The gates of the four MOSFETs are controlled by a switching control circuit similar to those seen in earlier chapters. The switching control circuit switches the MOSFETs on and off in pairs: Q1 with Q4, and Q2 with Q3. When one pair is turned on, the other is turned off.

We will begin by assuming that Q1 and Q4 are turned on. Therefore current flows through the armature of the DC motor and through Q1 and Q4. We will also assume that at this time the motor is turning clockwise. That is, the motor is operating in quadrant 1 and is now motoring.

How Does the Motor Operate in Quadrant 2?

Look at Figure 7-2 and assume that we now turn off Q1 and Q4. The armature of the motor will continue to rotate within the now collapsing magnetic field. As this rotation continues, magnetic lines of force are cut by the armature. This produces a voltage that opposes the applied voltage. This voltage will cause a current to flow from the armature, through CR4, to the supply, through CR1, and back to the armature. Since this voltage is opposite to the applied voltage, a counter torque is produced. This counter torque provides a braking action to

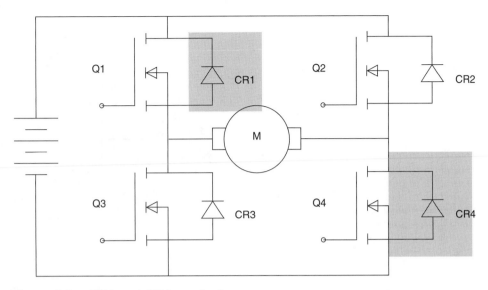

Figure 7-2: *CR1 and CR4 conducting.*

the motor. Notice, however, that the generated energy is now being applied to the power source. This is known as regenerative braking, and the motor is now operating in quadrant 2.

How Does the Motor Operate in Quadrant 3?

A motor operating in quadrant 3, as shown in Figure 7-3, must rotate in the opposite direction from a motor operating in quadrant 1. In our example, therefore, the motor now rotates counterclockwise. Turning on Q2 and Q3 in this case allows current flow in the reverse direction compared to quadrant 1, and this is what causes the motor to turn counterclockwise when it operates in quadrant 3.

How Does the Motor Operate in Quadrant 4?

Let us now turn off Q2 and Q3, as shown in Figure 7-4. As before, the armature of the motor will continue to rotate in the reverse direction within the now collapsing magnetic field. As this rotation continues, magnetic lines of force are cut by the armature. This produces a voltage that opposes the applied voltage. This voltage will cause a current to flow from the armature, through CR3, to the supply, through CR2, and back to the armature. Since this voltage is opposite to the applied voltage, a counter torque is produced. This counter torque provides a braking action to the motor. Notice, however, that the generated energy is now

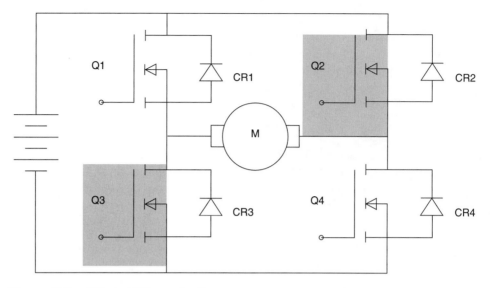

Figure 7-3: Q2 and Q3 conducting.

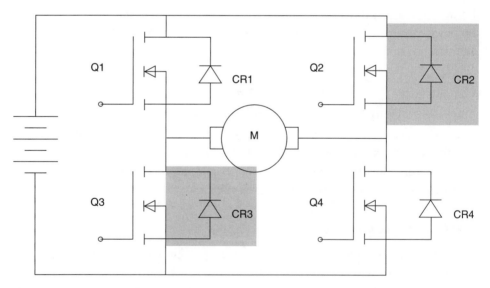

Figure 7-4: Q4 and CR3 conducting.

being applied to the power source. This provides regenerative braking from the reverse direction. The motor is now operating in quadrant 4.

How Can We Control the Motor's Speed in the Four Quadrants?

We can control a motor's speed quite easily in all four quadrants. Remember that each MOSFET is controlled by a pulse width modulation signal applied to its gate. If the width of the "on" pulse is wider, the MOSFET will conduct for a longer period of time, and the motor will operate at a higher average speed. If we make the "on" pulse narrower, the MOSFET will conduct for a shorter period of time, and the average speed of the motor will be lower. Varying the frequency of the control signal has no effect on the level of output voltage and current. However, a higher frequency control signal produces less ripple frequency in the output voltage and current, resulting in smoother motor operation. Therefore a four-quadrant chopper controls not only the quadrant of operation, but also the speed of the motor.

Review Questions

1. Describe each of the four quadrants as they affect motor operation.
2. Explain the term *regenerative braking*.
3. Explain the term *pulse width modulation* (PWM).
4. **True or False?** Motor speed increases as the "on" pulse becomes wider. Justify your answer.
5. **True or False?** Increasing the frequency of the "on" pulse will cause the motor's speed to increase. Justify your answer.
6. **True or False?** Increasing the frequency of the "on" pulse will cause the motor to operate more smoothly. Justify your answer.

Challenge Yourself

Refer to Figures D-3 and D-4 in Appendix D as you answer these questions.

1. What occurs at pin 2 of velocity integrator A1-A in Figure D-4?

2. Can this DC drive use CEMF as well as a tachometer-generator to indicate motor performance?

3. What are the functions of IC3-A and IC3-B in Figure D-3?

4. What are the functions of transformers T1, T2, T3, and T4 in Figure D-3?

Chapter 8

Troubleshooting DC Drives

OBJECTIVES

After completing this chapter, you will be able to:

- List the four main areas to check when troubleshooting a DC drive system.

- Discuss several safety considerations when troubleshooting a DC drive system.

- Define the term "phase imbalance."

- Calculate phase imbalance.

- List several possible causes of an inoperable DC drive system.

- List several types of adjustments found on DC drives.

- Explain the purpose of various adjustments found on DC drives.

In this chapter, we will discuss troubleshooting with respect to DC drives, identify the four main areas of possible problems, and present techniques for accurately and rapidly fixing these problems.

All too often, a maintenance technician who is introduced to something new will automatically assume that any problems are located in the new item. As we will see, this is not always true.

If the Motor Doesn't Run, Must the Problem Be in the Drive?

First, note that in a DC drive system there are four main areas to check for possible problems:

1. the electrical supply to the motor and the drive.

2. the motor and/or its load.

3. the feedback device and/or the sensors that provide signals to the drive.

4. the drive itself.

Even though the problem may be in any one or several of these areas, the best place to begin troubleshooting is at the drive unit itself. The reason for this is that most drives have some type of display that aids in troubleshooting. This display may be simply an LED that illuminates to indicate a specific fault condition, or an error or fault code that may be deciphered by looking it up in the operator's manual. For this reason, **it is strongly suggested** that a copy of the fault codes be made and fastened to the inside of the drive cabinet, where it will be readily accessible to the maintenance technician. The original should be placed in a safe location such as the maintenance supervisor's office.

After Checking the Drive and Reading the Code, It's Time to Make the Repairs

Wrong! Before going any further, let's take a minute to discuss safety. First, *stop and think about what you are doing!* Before working on any electrical circuit, *disconnect all power.* At times when this is not possible or permissible, *work carefully and wear the appropriate safety equipment.* Do not rely on safety interlocks, fuses, or circuit breakers to provide personnel protection. *Always use a voltmeter to verify that the equipment is de-energized, and tag and lock out the circuit!*

Even when the power has been disconnected, you are still subject to shock and burn hazards. Most drives have high-power resistors inside, and these can and do get *hot!* Give them time to cool down before touching them. Most drives also have large electrolytic capacitors that can and do store an electrical charge. Usually the capacitors have a bleeder circuit to dissipate this charge. However, be aware that this circuit may have failed. *Always verify that electrolytic capacitors are fully discharged* by carefully measuring any voltage present across the capacitor terminals. If voltage is present, *use an approved shorting device to discharge the capacitor completely.*

Having followed these steps, what should you do next? Use your senses! Most problems can be identified by using your senses.

Look! Do you see any charred or blackened components? Have you noticed any arcing? Do fuses or circuit breakers appear to have been blown or tripped? Do you see any discoloration around wires, terminals, or components? A good visual inspection can save a lot of troubleshooting time.

Listen! Did you hear any funny or unusual noises? A "frying" or "buzzing" sound may indicate arcing. A "hum" may be normal or an indication of loose laminations in a transformer core. A "rubbing" or "chafing" sound may indicate that a cooling fan is not rotating freely.

Smell! Do you notice any unusual odors? Burnt components and wires give off a distinctive odor when overheated. Metal will smell hot if there is too much friction.

Touch! (But very carefully!) If components feel cool, perhaps no current is flowing through the device. If components feel warm, chances are that everything is normal. If components feel hot, everything may be normal too, although it is more likely that the current is too high or the cooling device is not working properly. In any event there may be a problem worth further investigation.

The point is that by being *observant,* you have a good chance of discovering the problem or problem area. We will now discuss in more detail the four main areas mentioned earlier and some of the problems that we may encounter in each.

The Electrical Supply to the Motor and the Drive

Most maintenance technicians believe that the power distribution in an industrial environment is reliable, stable, and free of interference. Nothing could be further from reality! Power outages, voltage spikes and sags, and electrical noise are frequent occurrences. The effect of these phenomena is not as detrimental to motor performance as it can be to the operation of the drive itself. Most DC drives are designed to operate within a range of variation of supply voltages. Typically the incoming power can vary as much as ±10% with no noticeable change in drive performance. However, in the real world it is not unusual for power line fluctuations to exceed ten percent. These fluctuations may occasionally cause a controller to "trip," and if they occur repeatedly, a power line regulator may be required to hold the power at a constant level.

A power line regulator will be of little use, however, should the power supply to the controller fail. In this situation an **uninterruptable power supply (UPS)** will be needed. Several manufacturers produce complete power line conditioning units that combine a UPS with a power line regulator.

Quite often controllers are connected to an inappropriate supply voltage. For example, it is not unusual to find a drive rated 208 V connected to a 240 V supply. Likewise, a drive rated 440 V may be connected to a 460 V or even a 480 V source. Usually the source voltage should not exceed the voltage rating of the drive by more than 10%. For a drive rated 208 V the maximum supply voltage should not exceed 229 V (208 × 10% = 20.8 + 208 = 229).

Obviously, a 208 V drive connected to a 240 V power supply is overvoltaged and should not be used under such circumstances. For a 440 V drive the maximum supply voltage should not exceed 484 V (440 × 10% = 44 + 440 = 484). Furthermore, although this value is within acceptable limits, a potential problem still exists. Suppose that the power line voltage fluctuates by 10%. If the 440 V source suffers a 10% spike, the voltage will increase to 484 V. This value falls within the permissible design parameters of the drive. But consider what can happen if we connect a 440 V drive to a 460 V or 480 V power line. If we experience that same 10% spike, the voltage will increase to 506 V in the 460 V line, (460 × 10% = 46 + 460 = 506) and to 528 V in the 480 V line (480 × 10% = 48 + 480 = 528)!

Exceeding the voltage rating of the drive to this extent will probably damage some internal drive components. Most susceptible to excessive voltage and spikes or transients are the SCRs, MOSFETs, and power transistors. Premature failure of capacitors can also occur. As you can see, it is very important to match the line voltage to the voltage rating of the drive.

An equally serious problem occurs when the phase voltages are unbalanced. Typically, during construction care is taken to balance the electrical loads on the individual phases. As time goes by and new construction and remodeling occur, it is not unusual for the loading to become unbalanced. This imbalance will cause intermittent tripping of the controller, which can result in premature failure of certain components.

To determine if phase imbalance exists you must do the following:

1. Measure and record the **phase voltages** (L1 to L2, L2 to L3, and L1 to L3).

2. Add the three voltage measurements obtained in step 1 and record the *sum of the phase voltages.*

3. Dividing the sum obtained in step 2 by 3 and record the resulting *average phase voltage.*

4. Now subtract the average phase voltage obtained in step 3 from each phase voltage measurement taken in step 1 and record the results. (Treat any negative answers as positive values.) These values are the *individual phase imbalances.*

5. Add the individual phase imbalances obtained in step 4 and record the resulting *total phase imbalance.*

6. Divide the total phase imbalance obtained in step 5 by 2 and record the *adjusted total phase imbalance.*

7. Next, divide the adjusted total phase imbalance from step 6 by the average phase voltage found in step 3, and record the resulting *calculated phase imbalance.*

8. Finally, multiply the calculated phase imbalance from step 7 by 100 and record this *percent of total phase imbalance.*

Let's work through an example involving a 440 V three-phase supply to a DC drive to see how this procedure works.

1. Assume that L1 to L2 = 437 V, L2 to L3 = 443 V, and L1 to L3 = 444 V.

2. The sum of these phase voltages equals 437 V + 443 V + 444 V, or 1324 V.

3. The average phase voltage equals 1324 V ÷ 3, or 441.3 V.

4. To find the individual phase imbalances, we subtract the average phase voltage from the individual phase voltages and treat any negative values as positive. Therefore L1 to L2 = 437 V – 441.3 V, or 4.3 V; L2 to L3 = 443 V – 441.3 V, or 1.7 V; and L1 to L3 = 444 V – 441.3 V, or 2.7 V.

5. Now we find the total phase imbalance by adding together these individual phase imbalances: 4.3 V + 1.7 V + 2.7 V = 8.7 V.

6. To find the adjusted total phase imbalance, we divide the total phase imbalance by 2: 8.7 V ÷ 2 = 4.35 V.

7. Next, we divide the adjusted total phase imbalance by the average phase voltage to find the calculated phase imbalance: 4.35 V ÷ 441.3 V = 0.0099 V.

8. Finally, we multiply the calculated phase imbalance by 100 to find the percent total phase imbalance: 0.0099 × 100 = 0.99%.

In this example we are within tolerances and the differences in the phase voltages should not cause any problems. In fact, as long as the percent total phase imbalance does not exceed two percent, we should not experience any difficulties as a result of the differences in phase voltages.

What Should We Check When We Suspect That the Motor or the Load Is the Problem?

Probably the most common cause of motor failure is *Heat!* Excess heat can be simply a result of the motor's operating environment. Many motors are operated in areas of high ambient temperature. If steps are not taken to keep the motor within its operating temperature limits, the motor will ultimately fail.

Some motors have an internal fan to provide cooling. If such a motor is operated at reduced speed, the internal fan may not turn fast enough to cool the motor sufficiently. In these instances, an external fan may be needed to provide additional cooling to the motor. Typically, these fans are interlocked with the motor operation in such a way that the motor will not operate unless the fan operates as well. Therefore it is possible for a fault in an external fan control to prevent a motor from operating.

The temperature sensors used in motors generally consist simply of a non-adjustable thermostatic switch that is normally closed and opens only when the temperature rises beyond an acceptable level. Therefore it may be necessary to wait for an overheated motor to cool down sufficiently before the temperature sensor can be reset and the motor restarted.

Periodic inspection of the motor and any external cooling fans is strongly recommended. The fans should be checked for missing or bent vanes. All openings in the motor's and fan's housing intended to promote cooling should be kept free of obstructions. Any accumulation of dirt, grease, or oil there or elsewhere should be removed. Any filters used in the motor or fan must be cleaned or replaced on a routine schedule.

Heat may also cause other problems. When motor windings become overheated, the insulation on the wires may break down, causing a short circuit that may lead to an "open" condition. A common practice used to find shorts or opens in motor windings is to **megger** the windings with a **megohmmeter.** Extreme caution must be taken when using a megger. *When using a megger on the motor leads, be certain that you have disconnected the leads from the drive.* Failure to do so will cause the megger to apply a high voltage into the output section of the drive. Damage to the power semiconductors will result. You may decide to megger the motor leads at the drive cabinet. Again, be certain that you have disconnected the leads from the drive unit, and megger the motor leads and motor winding. *Never megger the output of the drive itself!*

Because the motor drives a load of some kind, the load may also create problems in motor operation. The drive may trip out if the load causes the motor to draw an excessive amount of current for too long a time. Most drives display some type of fault indication when this phenomenon occurs. This problem may be a result of excessive motor operating speed. Quite often, a minor reduction in speed is all that is necessary to prevent the repetitive tripping of the drive. The same effect occurs if the motor is truly overloaded by too large a load. Obviously, in this case either the motor size must be increased or the size of the load decreased to prevent the drive from tripping.

Some loads have a high inertia. They require not only a lot of energy to move them, but once moving, a lot of energy to stop them. If the drive cannot provide sufficient braking action to match the inertia of the load, the drive may trip, or

overhaul. A drive with greater braking capacity is needed to prevent the occurrence of overhauling.

What Problems Are Associated with Feedback Devices and Sensors?

Mechanical vibration may loosen the mounting or alter the alignment of feedback devices. Periodic inspections are necessary to verify that these devices are aligned and mounted properly.

It is also important to verify that the wiring to these devices is in good condition and that the terminations are clean and tight. Another consideration related to the wiring of feedback devices is electrical interference. Feedback devices produce low-voltage, low-current signals that are applied to the drive. If the signal wires from these devices are routed next to high-power cables, interference can occur. This interference may result in improper drive operation. To eliminate the possibility of interference, you must follow several steps. First, make certain that the signal wires from the feedback device are installed in a separate conduit. Do not install power wiring and signal wiring in the same conduit. The signal wires should consist of shielded cable, and the shield wire should be properly grounded at the drive cabinet only. Do not ground both ends of a shielded cable. When routing the shielded signal cable to its terminals in the drive cabinet, do not run or bundle the signal cable parallel to any power cables. The signal cable should be routed at a right angle to power cables. Also, do not route the signal cable near any high-power contactors or relays. When the coils of a contactor or relay are energized and de-energized, a spike is produced. This spike can also create interference in the drive. To suppress this spike, it may be necessary to install a free-wheeling diode across the DC coils, or a snubber circuit across the AC coils.

What Problems Exist in the Drive Itself?

First, look for any fault codes or fault indicators. Most drives provide some form of diagnostics, and this can be a great time-saver. The operator's manual interprets the fault codes and gives instructions for clearing the fault condition. However, there are other problems related to the drive that we should be aware of.

Heat can produce problems in the drive unit as well as elsewhere. The drive cabinet may have one or more cooling fans, with or without filters. These fans are often interlocked with the drive power in such a way that the fan must operate in order for the drive to operate. Make certain that the fans are operational and the filters are cleaned or replaced regularly. The drive's power semiconductors are typically mounted on heatsinks. A small thermostat may be mounted on a heatsink to detect excessive temperatures in the power semiconductors. If the heatsink becomes too hot, the thermostat opens, and the drive trips. Usually these thermostats reset themselves. You must wait for them to cool down and reset themselves before the drive will operate. If overheating occurs repeatedly, a more serious problem exists that will require further investigation.

If the drive is newly installed, problems are often the result of improper adjustments. On the other hand, if the drive has been in operation for some time, it is unlikely that readjustments are needed. All too often, an untrained individual will try to readjust a setting to "see if this fixes it!" Usually such readjustments only make things worse! This is not to say that adjustments are *never* needed. For example, changes in the process being controlled or replacement of some component of the equipment will probably require changes in the drive settings. It is therefore very important to record the current drive settings and any changes made to them over the years. This record should be placed in a safe location, and a copy of it should be made and placed in the drive cabinet for easy access by maintenance personnel.

Consider the more common adjustments that may be performed on typical DC drives. Not all drives permit all of these adjustments. Some drives require small adjustments performed with a screwdriver, whereas others allow you to program adjustments using a keypad or jumpers.

- *Acceleration / deceleration rates, or ramp:* This adjustment controls how rapidly the motor speeds up or slows down, as shown in Figure 8-1. If the motor must respond more quickly than its given load allows, the drive will trip.

- *Field voltage:* This adjustment sets the field voltage when the motor is not running. It not only saves energy, but also lowers the winding temperature and increases motor life expectancy.

- *IR compensation:* Used to sense CEMF from the motor as an indication of motor speed, this adjustment matches the motor's characteristics to the drive. Therefore readjustment should not be needed unless the motor or the drive is replaced. This adjustment is more commonly found on older drives and is rarely seen on drives that use feedback devices to sense motor speed.

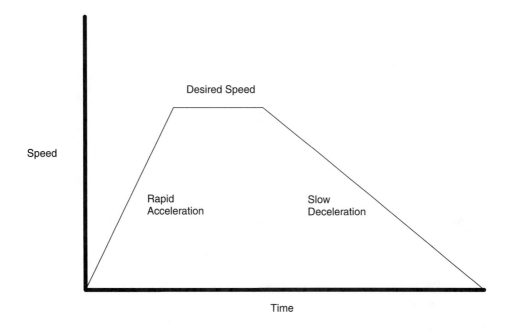

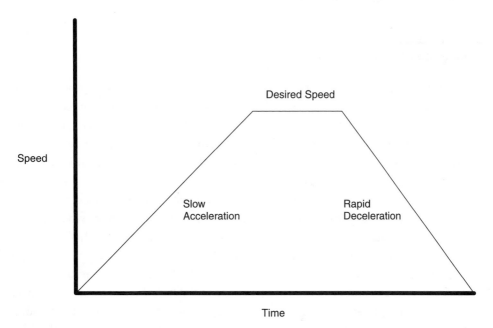

Figure 8-1: *Variations in acceleration and deceleration rates.*

- *Jogging or inching speed:* This is the speed of the motor expressed in small increments, usually 10% of the motor's full speed.

- *Maximum current:* This setting allows the motor to draw 150–300% more current than the motor's maximum rating for a short period of time. The higher the maximum current setting, the shorter is the period of time that the motor can draw this current.

- *Minimum current:* This setting prevents the DC motor from over-speeding in the event that the shunt field circuit opens or the shunt field current becomes too small to produce a magnetic field strong enough to generate sufficient torque.

- *Overspeeding:* Typically, this setting trips the drive if the motor speed exceeds the desired speed by more than 10%.

- *Watchdog circuit:* Adjusted to detect certain levels of electrical interference or "noise," this setting also trips the drive in the event of a voltage sag, spike, or single phasing.

Remember, you should *never* adjust these settings unless you have been properly trained and know their effects.

What If the Drive Still Doesn't Work?

Possibly you need a new one. Replacing the drive should clearly be the last choice. More often than not, even if the drive is defective, something external to the drive is the reason for the drive's failure. Replacing the drive without determining the cause of its failure may cause damage to the replacement drive. However, whenever a drive fails, regardless of the cause, some possibility exists that you can get it to work again.

Most drive failures occur in the power section, where you will find the power SCRs, transistors, MOSFETs, and so on. In some drives, these devices will be individual components. You can test these devices with reasonable accuracy using nothing more than an ohmmeter. (See Appendix C.) In other drives, the SCRs, transistors, and MOSFETs are contained within a power module.

An SCR power module is shown in Figure 8-2. One of these modules would be used for each phase of the three-phase AC. Notice that this module contains two SCRs. We will call the SCR to the left "A" and the SCR to the right "B." You can use an ohmmeter to test an SCR module with reasonable accuracy. To do so, you must understand what each terminal of the module represents. A schematic diagram of the SCR module is shown in Figure 8-3.

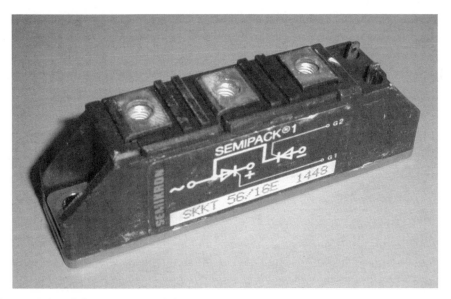

Figure 8-2: *SCR power module.*

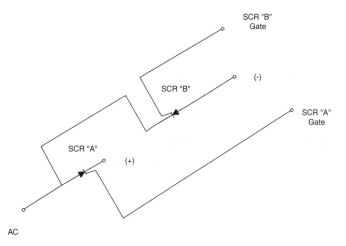

Figure 8-3: *SCR power module schematic.*

Notice in Figure 8-2 that there are a total of five terminals on the SCR module. Referring to Figure 8-3, you will see the same five terminals. Two of the terminals (the small terminals in Figure 8-2), represent the gate connections for each SCR. The gate terminal toward the back of the module is connected to the gate of SCR "B," while the gate terminal at the front of the module is connected to the gate of SCR "A." The large terminal closest to the gate terminals is the negative terminal of the module, which is also the anode of SCR "B." The large terminal farthest from the gate terminals is the AC terminal for the module, which is also the anode of SCR "A." Notice that the anode of SCR "A" is internally connected to the cathode of SCR "B." The large terminal in the center is the positive terminal of the module, which is also the cathode of SCR "A."

To test the module with an ohmmeter, follow these steps:

1. Place the ohmmeter in the "Ohms" position, if using a digital multimeters. (Use the 200 Ω position if using an analog multimeters.)

2. Place the black (negative) lead of your ohmmeter on the anode (terminal farthest from the gate terminals) of SCR "A."

3. Place the red (positive) lead of your ohmmeter on the cathode (center terminal) of SCR "A."

4. Your meter should indicate a very high or infinite resistance. A low resistance reading indicates a faulty SCR. Replace the module.

5. Reverse your meter connections. Place the black (negative) lead of your ohmmeter on the cathode (center terminal) of SCR "A."

6. Place the red (positive) lead of your ohmmeter on the anode (terminal farthest from the gate terminals) of SCR "A."

7. Your meter should indicate a very high or infinite resistance. A low resistance reading indicates a faulty SCR. Replace the module.

8. While leaving your ohmmeter connected as in steps 5 and 6 above, use a clip lead to connect the gate of SCR "A" (small terminal at the front of the module) to the red (positive) lead of your ohmmeter.

9. You should notice a drop in the resistance reading. This is a result of your triggering SCR "A" into conduction. If your resistance reading does not drop, the SCR may be faulty and you should replace the module. However, it is possible that your ohmmeter is not supplying sufficient current to trigger the SCR into conduction, and the SCR may be functioning normally. The old saying, "If in doubt, change it out!" would apply.

If SCR "A" appears normal, repeat the same process to check SCR "B":

1. Place the black (negative) lead of your ohmmeter on the anode (terminal closest to the gate terminals) of SCR "B."

2. Place the red (positive) lead of your ohmmeter on the cathode (terminal farthest from the gate terminals) of SCR "B."

3. Your meter should indicate a very high or infinite resistance. A low resistance reading indicates a faulty SCR. Replace the module.

4. Reverse your meter connections. Place the black (negative) lead of your ohmmeter on the cathode (terminal farthest from the gate terminals) of SCR "B."

5. Place the red (positive) lead of your ohmmeter on the anode (terminal closest to the gate terminals) of SCR "B."

6. Your meter should indicate a very high or infinite resistance. A low resistance reading indicates a faulty SCR. Replace the module.

7. While leaving your ohmmeter connected as in steps 5 and 6 above, use a clip lead to connect the gate of SCR "B" (small terminal at the back of the module) to the red (positive) lead of your ohmmeter.

8. You should notice a drop in the resistance reading. This is a result of your triggering SCR "B" into conduction. If your resistance reading does not drop, the SCR may be faulty and you should replace the module. However, it is possible that your ohmmeter is not supplying sufficient current to trigger the SCR into conduction, and the SCR may be functioning normally. Again, "If in doubt, change it out!"

While we are on the subject of power modules, there is another type of power module that you may find in the drive upon which you are working. This module is technically not part of the power output section, although it does have a power function in the drive. The module is a three-phase bridge rectifier module. It is used to convert three-phase AC into rectified DC. A picture of this module appears in Figure 8-4 and a schematic of this module appears in Figure 8-5.

Notice that this module has five terminals. The two horizontal terminals at the left end of the module are the (+) and (–) DC connections. The three vertical terminals are the connections for the three-phase AC (L1, L2, and L3).

To test the module with an ohmmeter, follow these steps:

1. Place the ohmmeter in the "Diode Test" position, if using a digital multimeters. (Use the 200 Ω position if using an analog multimeters.)

2. Place the black (negative) lead of your ohmmeter on the (–) terminal of the module.

3. Place the red (positive) lead of your ohmmeter on AC terminal "L1" (this is actually the cathode of diode "CR4").

4. Your meter should indicate a very high or infinite resistance. A low resistance reading indicates a faulty diode. Replace the module.

5. Move the red (positive) lead of your ohmmeter from AC terminal "L1" to AC terminal "L2" (this is actually the cathode of diode "CR5").

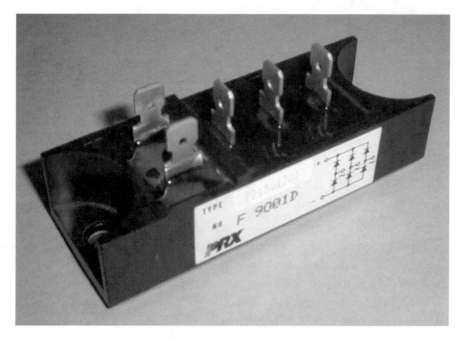

Figure 8-4:   *Three-phase bridge rectifier module.*

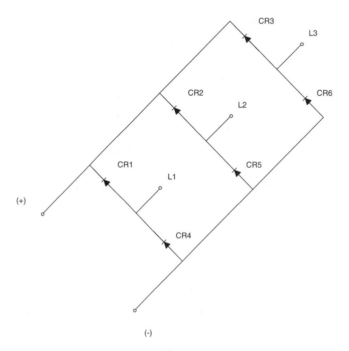

Figure 8-5:   *Three-phase bridge rectifier module schematic.*

6. Your meter should indicate a very high or infinite resistance. A low resistance reading indicates a faulty diode. Replace the module.

7. Move the red (positive) lead of your ohmmeter from AC terminal "L2" to AC terminal "L3" (this is actually the cathode of diode "CR6").

8. Your meter should indicate a very high or infinite resistance. A low resistance reading indicates a faulty diode. Replace the module.

9. Reverse your meter connections. Place the red (positive) lead of your ohmmeter on the (–) terminal of the module.

10. Place the black (negative) lead of your ohmmeter on AC terminal "L1" (this is actually the cathode of diode "CR4").

11. Your meter should indicate a low resistance. A high or infinite resistance reading indicates a faulty diode. Replace the module.

12. Move the black (negative) lead of your ohmmeter from AC terminal "L1" to AC terminal "L2" (this is actually the cathode of diode "CR5").

13. Your meter should indicate a low resistance. A high or infinite resistance reading indicates a faulty diode. Replace the module.

14. Move the black (negative) lead of your ohmmeter from AC terminal "L2" to AC terminal "L3" (this is actually the cathode of diode "CR6").

15. Your meter should indicate a low resistance. A high or infinite resistance reading indicates a faulty diode. Replace the module.

16. Place the red (positive) lead of your ohmmeter on the (+) terminal of the module.

17. Place the black (negative) lead of your ohmmeter on AC terminal "L1" (this is actually the anode of diode "CR1").

18. Your meter should indicate a very high or infinite resistance. A low resistance reading indicates a faulty diode. Replace the module.

19. Move the black (negative) lead of your ohmmeter from AC terminal "L1" to AC terminal "L2" (this is actually the anode of diode "CR2").

20. Your meter should indicate a very high or infinite resistance. A low resistance reading indicates a faulty diode. Replace the module.

21. Move the black (negative) lead of your ohmmeter from AC terminal "L2" to AC terminal "L3" (this is actually the anode of diode "CR3").

22. Your meter should indicate a very high or infinite resistance. A low resistance reading indicates a faulty diode. Replace the module.

23. Reverse your meter connections. Place the black (negative) lead of your ohmmeter on the (+) terminal of the module.

24. Place the red (positive) lead of your ohmmeter on AC terminal "L1" (this is actually the anode of diode "CR1").

25. Your meter should indicate a low resistance. A high or infinite resistance reading indicates a faulty diode. Replace the module.

26. Move the black (negative) lead of your ohmmeter from AC terminal "L1" to AC terminal "L2" (this is actually the anode of diode "CR2").

27. Your meter should indicate a low resistance. A high or infinite resistance reading indicates a faulty diode. Replace the module.

28. Move the black (negative) lead of your ohmmeter from AC terminal "L2" to AC terminal "L3" (this is actually the anode of diode "CR3").

29. Your meter should indicate a low resistance. A high or infinite resistance reading indicates a faulty diode. Replace the module.

Once you have completed testing the module, and you determine that the module is defective, you can usually obtain a substitute part from a local electronics parts supplier. If the part is not available, you will have to return the drive to the manufacturer for service or call a service technician for on-site repairs.

If the problem is not in the drive's power section, then it must be located in its control section. The electronics in the control section are more complex, and therefore troubleshooting is not recommended. In this event, you should definitely return the drive to the manufacturer for repair or call a service technician for on-site repairs.

Review Questions

1. Name the four main areas to check when a problem with a DC drive system occurs.

2. Where is the best place to begin troubleshooting a DC drive system problem? Why?

3. List some safety steps that you should follow prior to working on a DC drive system.

4. Is it permissible to connect a DC drive to a supply voltage that is higher than the nameplate rating of the drive? Why or why not?

5. Explain phase imbalance.

6. Calculate the phase imbalance for a supply to a DC drive that has the following voltages: L1 to L2 = 209 V; L2 to L3 = 205 V; and L1 to L3 = 210 V.

7. **True or False?** An inoperable cooling fan can prevent a drive from operating. Explain your answer.

8. Describe the cooling problem created by using a variable speed drive to operate a motor with a shaft-mounted fan.

9. Describe the dangers of using a "megger" on a DC drive system.

10. Explain how the motor load can affect the performance of a DC drive.

11. **True or False?** When using shielded cable on feedback devices, you must ground the shield wire at both ends of the cable. Justify your answer.

12. Describe any precautions that you should observe when routing the feedback device cable outside the drive cabinet.

13. Describe any precautions that you should observe when routing the feedback device cable inside the drive cabinet.

14. Explain when it is permissible to readjust drive settings.

15. Describe the purpose of the IR compensation adjustment.

16. Which two adjustments inhibit DC motor overspeeding?

17. Explain when a DC drive should be returned to the manufacturer for repair.

Chapter 9

AC Drive Fundamentals

OBJECTIVES

After completing this chapter, you will be able to:

- Discuss, in general, the purpose of an AC drive.

- Discuss, in general, the operating principle of an AC drive.

- List the three main sections of an AC drive.

For many years, the mainstay motor of industry has been the three-phase squirrel cage induction motor. This motor has the advantages of low cost and low maintenance. Its biggest disadvantage has been its fixed operating speed. If you needed a three-phase induction motor with variable speed, you had to use a wound-rotor induction motor with a potentiometer. This configuration entailed added expense and increased maintenance.

Fortunately, we now have AC drives, more commonly called inverters (Figure 9-1). Inverters are now used not only to vary the speed of the squirrel cage motor, but also to vary its torque, start the motor slowly and smoothly, and increase the motor's efficiency. In the following chapters, we will learn how inverters accomplish these functions. However, first we will become acquainted in this chapter with the structure of inverters.

An AC Drive has three basic sections: the converter, the DC filter, and the inverter. Figure 9-2 shows the inside view of an inverter drive. The **converter** rectifies the applied AC into DC. The **DC filter** (also called the DC link or DC bus) provides a smooth, rectified DC. The **inverter** switches the DC on and off so rapidly that the motor receives a pulsating DC that appears similar to AC. Because this DC switching is controlled, we can vary the frequency of the artificial AC that is applied to the motor, something that we are normally unable to do.

The AC line frequency is set by the electric utility companies across the United States at a standard rate of 60 Hz. Because the AC line frequency could not normally be varied, the speed of a squirrel cage motor was, for the most part, fixed. The number of the motor's poles could be increased or decreased, causing

Figure 9-1:   *Carotron Vista II digital inverter drive.*

the motor to slow down or speed up, respectively. However, this could only be done as the motor was being built. We could also vary the stator voltage. However, doing so reduces the motor's ability to drive a load at low speeds, so this method has limited usefulness.

Today, we can use inverters to vary the frequency of the "AC" (the pulsating DC) applied to the squirrel cage motor, and thus vary the motor's speed, while maintaining constant torque. As we will see, in today's inverter drives the stator voltage and the frequency are both variable.

Figure 9-2: *Inside view of a digital inverter drive.*

The different methods used to create the pulsating DC constitute the basic differences among the various types of inverter drives. Another area of design variation is that of the ratio of volts to hertz (**V/Hz**). The V/Hz ratio should be maintained at a constant value. This means that a motor turning at 1800 RPM, operating from 208 V at 60 Hz, would have to operate from 104 V at 30 Hz to attain a speed of 900 RPM because these values do not change the V/Hz ratio from its original value. We will learn more about this requirement in later chapters.

In the remaining chapters of this book we will also learn about the two major types of inverter drives: the **voltage source inverter,** also known as the **voltage fed inverter,** or **VSI,** and the **current source inverter.** The voltage source inverter may be further subdivided into the categories of the variable voltage inverter and the pulse-width modulated inverter.

Finally, we will end our general discussion of AC inverter drives with a chapter on the various types of adjustments and maintenance that we can perform on them. We will also identify areas of concern when troubleshooting an inoperable inverter drive.

Review Questions

1. **True or False?** Another name for an AC drive is a converter.

2. The DC filter is also known as

   a. the DC bus.

   b. the DC link.

   c. the converter.

   d. pulsating DC.

3. Explain, in general, the principles behind the operation of an AC inverter drive.

4. List the three basic sections of an AC inverter drive and describe the purpose of each section.

5. Explain why we need an AC inverter drive.

6. List two major types of inverter drives.

Chapter 10

Variable Voltage Inverters

OBJECTIVES

After completing this chapter, you will be able to:

- Discuss the operation of a variable voltage inverter.
- Discuss the meaning of the term V/Hz and its importance.
- Explain how phase control is achieved.
- Explain the six-step method of phase control.

In this chapter, we will learn about variable voltage inverter drives. We will study the six-step method, with phase control and with chopper control.

What Is a Variable Voltage Inverter?

A **variable voltage inverter (VVI)** is one of two categories of adjustable frequency, variable speed AC drives. In a VVI the DC voltage is controlled, and the DC current is free to respond to the motor needs. The VVI uses a converter, a DC link, and an inverter to vary the frequency of the applied AC voltage.

What Is a Converter?

A **converter** is a circuit that changes the incoming AC power (fixed voltage, fixed frequency) into DC power. The converter circuit is simply a rectifier circuit that produces an unfiltered, pulsating DC. The converter can be either single-phase or three-phase, depending on the type of power that the AC induction motor you are using needs.

What Is a DC Link?

A typical **DC link,** also called a DC filter or DC bus, is simply a filter circuit composed of an inductor and a capacitor. The purpose of the DC link is to filter or smooth the AC ripple from the output of the converter stage. The filtered DC from the DC link is fed via the DC bus to the input of the inverter stage.

What Is an Inverter?

An **inverter** converts the applied DC voltage to a pulsating DC voltage. Because we can vary the magnitude and the frequency of this pulsating DC, we can therefore control the speed of an AC induction motor. This pulsating DC power acts as artificial AC power on the induction motor.

How Does a VVI Operate?

The VVI shown in the schematic diagram in Figure 10-1 is divided into three basic parts: the converter, the DC link, and the inverter. Basically, three-phase AC is applied to the converter stage of the VVI. Naturally, this AC has both fixed amplitude and fixed frequency. The converter rectifies the AC into DC, which is then smoothed by the DC link. This filtered DC is applied to the inverter, which chops the DC into pulsating AC. This AC is then applied to the motor. The chopping rate applied to the DC varies the frequency of the AC applied to the motor.

We can look at this in more detail in Figures 10-2A through 10-2D (see pages 93 and 94), which are simplified schematics of the inverter stage of a six-step VVI. The inverter stage functions in the same way regardless of whether you use the phase control or the chopper control method, both of which we will study later In the first sixth of the cycle, transistors Q1, Q5, and Q6 conduct. In the second sixth of the cycle, transistors Q1 and Q6 continue to conduct; however,

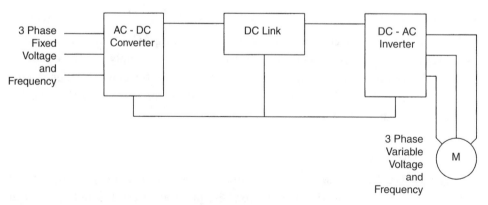

Figure 10-1:   *Variable voltage inverter block diagram.*

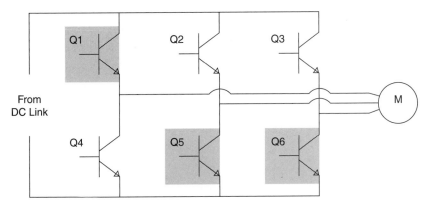

Figure 10-2A: *Six-step inverter with Q1, Q5, and Q6 conducting.*

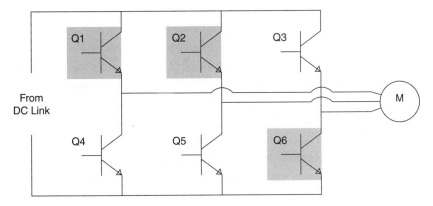

Figure 10-2B: *Six-step inverter with Q1, Q2, and Q6 conducting.*

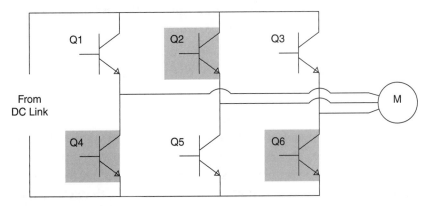

Figure 10-2C: *Six-step inverter with Q2, Q4, and Q6 conducting.*

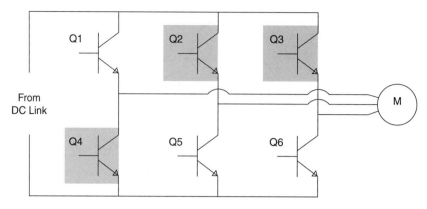

Figure 10-2D: *Six-step inverter with Q2, Q3, and Q4 conducting.*

transistor Q5 turns off, and transistor Q2 turns on. During the third sixth of the cycle, transistors Q2 and Q6 continue to conduct, but transistor Q1 turns off, and transistor Q4 conducts. During the fourth sixth of the cycle, transistors Q2 and Q4 remain on, while transistor Q6 turns off and Q3 turns on. After two more steps, the motor will have completed one revolution. This six-step cycle is the reason for the inverter's name.

By turning the transistors on at a slower or faster rate, we can vary the frequency of the voltage applied to the motor. As the frequency varies, the inductive reactance of the motor windings also varies proportionally. An increase in reactance causes a decrease in current at higher speeds, and as a result the motor will not function properly. Therefore the voltage must be increased proportionally to the frequency, in a ratio known as the volts/hertz ratio, or V/Hz. The V/Hz ratio must be kept constant, and this can be accomplished by one of two methods: phase control or chopper control.

What Is Phase Control?

In Figure 10-3 (see page 95), the converter stage is composed of a three-phase SCR bridge circuit. Remember that we can control when and for how long an SCR conducts. If we turn the SCR on early in the cycle, we can provide a higher average voltage to the inverter because the SCR will conduct for a longer portion of the cycle. If the SCR is turned on late in the cycle, the SCR will conduct for a shorter period of time, and therefore the inverter will receive a lower average voltage. By varying the DC bus voltage level in this way and simultaneously

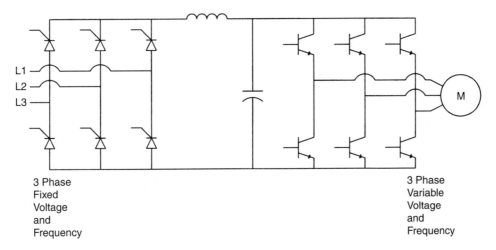

L1
L2
L3

3 Phase
Fixed
Voltage
and
Frequency

M

3 Phase
Variable
Voltage
and
Frequency

Figure 10-3: *Variable voltage inverter with phase control.*

varying the frequency of the output from the inverter section, we can maintain a constant V/Hz ratio.

Why Is the V/Hz Ratio Important?

The V/Hz ratio (also called the V/f ratio) is very important for a number of reasons. Remember that we need to be concerned with many characteristics when dealing with AC and inductors. We not only have to be aware of voltage, current, and resistance, but we must also recognize the importance of reactance, hysteresis, and eddy currents. We will see what can happen if we fail to maintain a constant V/Hz ratio.

Suppose that we could increase the voltage applied to the motor without adjusting the frequency of the applied voltage. What would happen? The increased voltage would produce increased magnetic flux, which in turn would saturate the iron components of the motor. This flux would cause increased iron losses in the form of hysteresis and eddy currents. It would also increase the stator current and possibly damage the motor windings as a result.

If we examined what happened to the ratio of volts to hertz (V/Hz) in the preceding example, we would see that the ratio increased because the frequency remained constant as the voltage increased. The same effect occurs if the voltage is kept constant as the frequency is reduced. Excessive current will flow which will result in more heat produced. As we saw in the example, the net results of such change to the V/Hz ratio are the disruption of normal motor operation and the potential for serious damage to the motor. That is why it is so important to continued and proper motor operation that we maintain a constant V/Hz ratio.

What Is Chopper Control?

In Figure 10-4 we begin again with the same inverter section discussed earlier. Here we have returned to a simple three-phase diode bridge rectifier circuit. However, we have now added a chopper, which is nothing more than an electronic switch. This switch may be a transistor or, more commonly, a MOSFET.

Here is how the chopper works. Three-phase AC is applied to the three-phase diode bridge. The bridge rectifies the AC into a fixed DC voltage. Capacitor C1 filters the DC for the chopper circuit. The smoothed DC is then "chopped" at a fixed frequency by electronic switches; however, the ratio of the chopper's "on" time to "off" time is varied. If the chopper is turned on for a longer period of time, a higher DC voltage is applied to the inverter. If the chopper is turned off for a longer period of time, the DC applied to the inverter will be lower. Varying these "on" and "off" times thus allows us to maintain a constant V/Hz ratio.

What Are the Advantages/Disadvantages of Each Type of Control?

All types of VVIs provide open-circuit protection, can handle multiple motor applications and undersized motors, and do not require high-speed switching

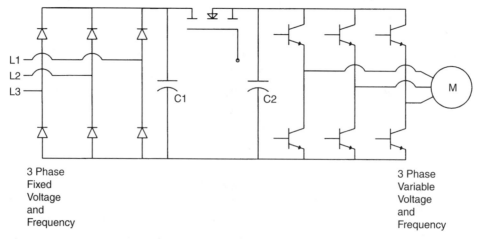

3 Phase
Fixed
Voltage
and
Frequency

3 Phase
Variable
Voltage
and
Frequency

Figure 10-4:   *Variable voltage inverter with chopper control.*

devices. These drives are quite light in weight, utilize relatively simple control circuits, and exhibit good efficiency at low speeds.

On the negative side, all VVIs lack short-circuit protection and they cannot handle oversized motors. Another disadvantage that both types of VVIs share is that they do not operate smoothly at low frequencies (< 6 Hz). Regenerative braking is not possible with a chopper-controlled VVI, but is possible with a phase-controlled VVI. On the other hand, a chopper-controlled VVI can be operated from batteries, whereas a phase-controlled VVI cannot.

Review Questions

1. What do the letters *VVI* mean?

2. **True or False?** In a VVI the DC voltage is controlled.

3. Describe the function of a converter.

4. Describe the function of an inverter.

5. Explain what the V/Hz ratio is and why it is so important.

6. Explain how phase control is accomplished.

7. A chopper

   a. changes AC into DC.

   b. changes DC into AC.

   c. changes pulsating DC into steady DC.

   d. changes steady DC into pulsating DC.

8. Which type of VVI, phase- or chopper-controlled, can be operated from batteries?

9. Is regenerative braking possible with a chopper-controlled VVI?

10. Name one disadvantage common to both types of VVI.

Chapter 11

Pulse Width Modulated Variable Voltage Inverters

OBJECTIVES

After completing this chapter, you will be able to:

- Discuss the similarities between a pulse width modulated VVI and the SCR armature voltage controller.

- Discuss how the firing angles of the SCRs are varied.

- Discuss the operation of a pulse width modulated VVI.

In this chapter, we discuss another form of variable voltage inverter that uses pulse width modulation (PWM) to vary the voltage applied to an AC induction motor. Many similarities exist between this type of inverter and the DC drives discussed in Chapters 2 and 3. Therefore, we will begin by reviewing briefly the theory behind DC drives to give you a better understanding of how a PWM VVI operates.

How Does the PWM VVI Sense Motor Speed?

Notice in Figure 11-1 on the next page that the feedback device used is a tachometer-generator. (You will find it located in the upper right corner of the schematic.) Recall that the tachometer-generator is a DC generator attached to an AC induction motor's shaft. As the AC induction motor turns, the DC tachometer-generator also turns, producing a positive DC voltage. The faster the AC induction motor turns, the more positive DC voltage the DC tachometer-generator produces. In other words, the output of the DC tachometer-generator is proportional to the speed of the AC induction motor.

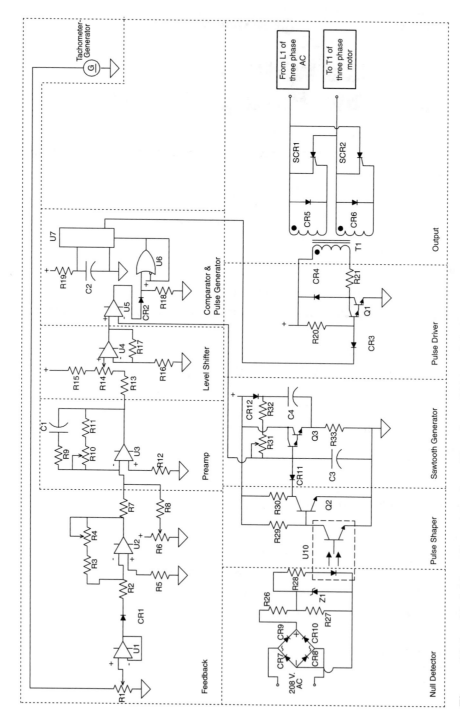

Figure 11-1: *Variable voltage inverter with pulse width modulation.*

What Does the Feedback Section Do?

Now we will consider the feedback section, shown in Figure 11-2 (see page 102). The positive voltage, or feedback signal, produced by the DC tachometer-generator is fed through R1 into buffer amplifier U1. The output of U1 is then fed to the inverting input of U2, which inverts the polarity of the tachometer-generator voltage from positive to negative. The resulting negative voltage is then fed through R7 to the noninverting input of U3 in the preamplifier section. Notice that one end of R8 is also connected to the noninverting input of U3 and that the other end of R8 is connected to variable resistor R6. Resistor R6 allows us to set the reference voltage level, sometimes called the command signal, which is the voltage necessary for the AC induction motor to run at a certain speed. The reference voltage (positive) and the tachometer-generator voltage (negative) are added together at the junction of R7 and R8, which is known as the summing point. The difference of these two voltages is the value of the voltage that will appear at the noninverting input of U3. This value is known as the error signal. Now let's leave this section for a moment, and move on to the null detector and sawtooth generator circuits.

What Do the Null Detector and the Sawtooth Generator Do?

Notice that in the null detector shown in Figure 11-3 the applied AC voltage is rectified by the full-wave bridge circuit consisting of CR7–CR10. This process causes an unfiltered, pulsating DC voltage to appear across Zener diode Z1 and also to appear across R28 and the LED of optocoupler U10.

What happens as the pulsating DC voltage rises from 0 V to its peak value? The LED of U10 does not conduct until the DC voltage reaches the turn-on voltage of the LED. Assume that this value is approximately 1.5 V. As the DC voltage increases from 0 V to 1.5 V, U10 is off. Therefore the output of U10 is positive, causing Q2 to be on. When Q2 conducts, a negative pulse appears at the collector of Q2. This negative pulse is used to reset the output of the sawtooth generator back to 0.

What happens when the LED of U10 turns on? As the pulsating DC voltage rises, eventually it will reach 1.5 V. The voltage across the LED of U10 will never exceed the voltage rating of Zener diode Z1. At this point the LED of U10 turns on and conducts, causing the output of U10 to decrease to a level that turns Q2 off. With Q2 turned off, the sawtooth generator outputs a sawtooth waveform to the inverting input of U5 (the comparator). The sawtooth generator continues to output this waveform until it is reset when Q2 turns on, that

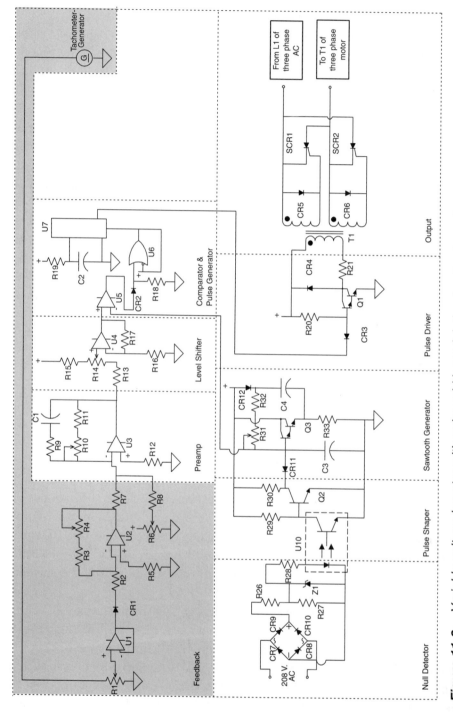

Figure 11-2: Variable voltage inverter with pulse width modulation feedback section.

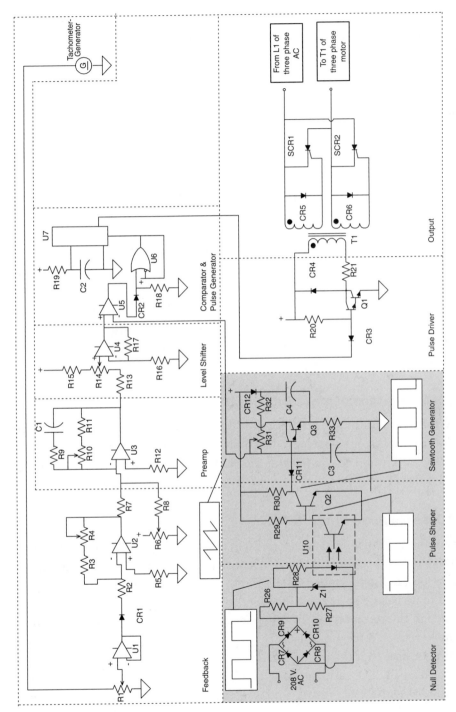

Figure 11-3: *Variable voltage inverter with pulse width modulation null detector and sawtooth generator section.*

is, when the pulsating DC voltage falls below 1.5 V and decreases to 0 V. Therefore, the pulsating DC voltage causes the sawtooth generator to provide a reference signal to the comparator. This reference signal is used to vary the firing angle of the SCRs that control the stator voltage.

What Does the Comparator Do?

Consider U5, the comparator, in Figure 11-4. The comparator has two inputs. One comes from the feedback circuit and is a measure of the speed of the armature. The other input, as we just learned, is a ramp signal from the sawtooth generator. The noninverting input from the feedback amplifier will be constant if the load on the armature is constant. In contrast, the inverting input is a rising amplitude signal from the sawtooth generator. Initially, the noninverting input will cause the output of U5 to be positive. However, as the inverting input ramp signal climbs, a point is reached where the amplitudes of both inputs are equal. At this point, the output of U5 will switch off, becoming a rectangular pulse. The width of this pulse is determined by the length of time it takes for both inputs to become equal. This is what determines whether the SCRs conduct earlier or later within each half cycle. This process is similar to that of the comparator stage discussed in Chapter 3.

What Is the Function of the Pulse Generator?

The Pulse generator shown in Figure 11-5 (see page 106) is an integrated circuit that contains a one-shot, or monostable-multivibrator. Its purpose is to trigger on the output of the comparator and in turn provide a narrow pulse to Q1, the pulse driver.

What Does the Pulse Driver Do?

Referring to Figure 11-6 (see page 107), recall that Q1 does not conduct until a positive pulse appears at its base. When this occurs, current flows through the primary of T1. Furthermore Q1 conducts only for the duration of the pulse applied to its base. The current flowing through the primary of T1 causes a current flow in the secondary of T1. Therefore the gates of SCR1 and SCR2 simultaneously receive a trigger pulse.

However, note that the anode of SCR1 and the cathode of SCR2 are connected to motor lead T1. The cathode of SCR1 and the anode of SCR2 are connected to L1 of the AC source. As a result, when the cathode of SCR1 is positive, the anode of SCR2 is positive. When the cathode of SCR1 is negative, the anode of SCR2

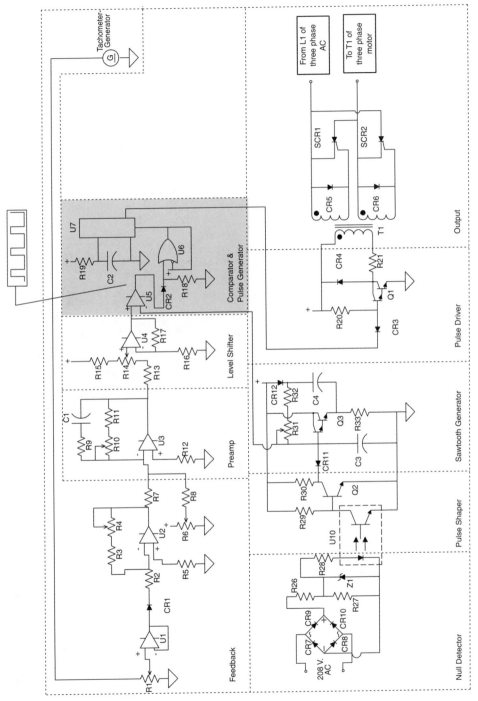

Figure 11-4: *Variable voltage inverter with pulse width modulation comparator section.*

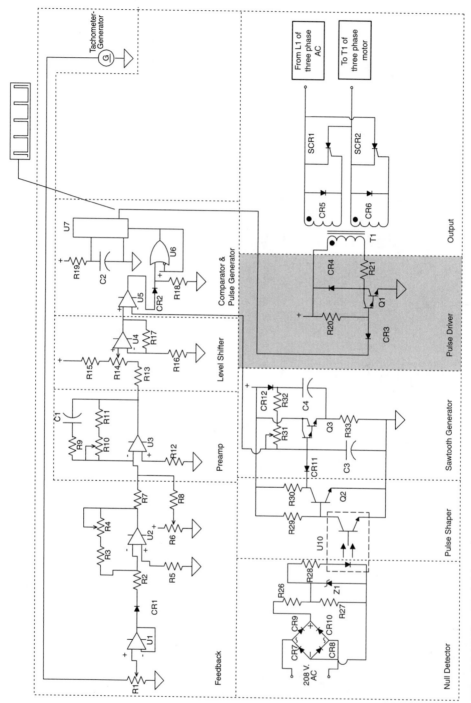

Figure 11-5: Variable voltage inverter with pulse width modulation pulse generator section.

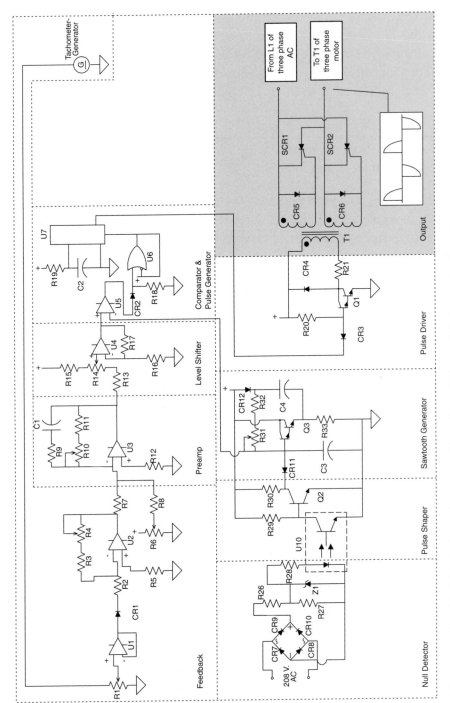

Figure 11-6: Variable voltage inverter with pulse width modulation output section.

is negative. Therefore, when SCR1 and SCR2 are triggered, only the SCR with the positive anode or negative cathode conducts. This causes the SCRs to conduct alternately. Thus AC current flows through the stator of the AC induction motor. For simplicity's sake, we have considered only one phase of a three-phase controller here. The same circuitry is duplicated for each phase.

How Does This Process Automatically Vary an AC Induction Motor's Speed?

Now let's bring this all together. Figure 11-7 shows the waveforms at various points in the comparator and Pulse Generator, Pulse Driver, and Output stages. These waveforms have been lined up vertically so that you understand the timing of the waveforms that must occur in order for the output SCRs to trigger at the appropriate time. Here's how it works. The waveform in Figure 11-7a shows the sawtooth waveform at the inverting input of U5, the comparator. At the same time, Figure 11-7a shows the DC voltage that is applied to the non-inverting input of comparator U5. Recall that this voltage is the feedback voltage from the tachometer-generator, and is an indication of the armature speed. The DC voltage level shown is representative of a motor speed of 1800 RPM. These two inputs cause the output of the comparator to produce a pulsating DC as shown in Figure 11-7b. Notice that the pulse is initially positive, and is switched off when the two inputs of the comparator become equal. The output of U5 remains off until the sawtooth generator resets. At this point, the output of U5 switches on, and remains on until the sawtooth signal is equal to the feedback voltage.

The pulsating output of U5 is fed through diode CR2 to the input of U6. Because of CR2, the input signal to U6 will be inverted compared to the output of U5. This is shown in Figure 11-7c. The positive pulses from the output of U6 trigger the one-shot, U7. The output of the one-shot, U7, will appear in phase with the input pulses as seen in Figure 11-7d.

The output of the one-shot, U7, is fed through diode CR3 to the base of the Darlington transistor, Q1. Notice in Figure 11-7e that the collector of Q1 is in phase with the output of the one-shot. When Q1 conducts, current will flow through the pulse transformer, T1.

When current flows through pulse transformer T1, a pulse appears at the gates of SCR1 and SCR2. This is shown in Figure 11-7f. Recall that only the SCR that is properly biased will conduct. The SCR that is triggered and properly biased will conduct until the AC voltage drops to the zero crossing point.

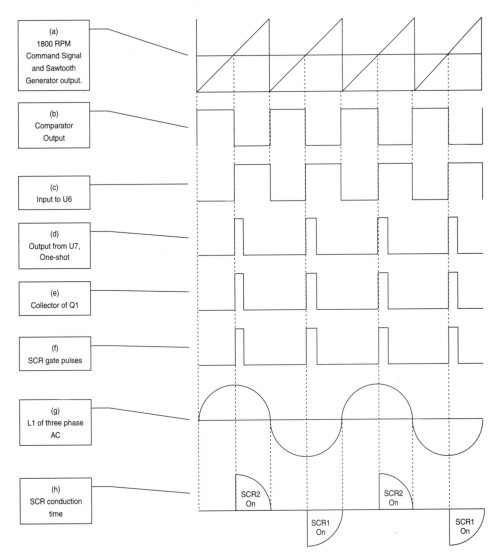

Figure 11-7: *Waveform timing and SCR conduction time at 1800 RPM.*

(The AC voltage waveform is shown in Figure 11-7g.) The next pulse from T1 will cause the other SCR to conduct. Therefore, the SCRs will conduct alternately as shown in Figure 11-7h.

Let's assume that an increase in load has caused the motor speed to decrease to 1000 RPM. Refer to Figure 11-8. The waveform in Figure 11-8a shows the sawtooth waveform at the inverting input of U5, the comparator. At the same time, Figure 11-8a also shows the DC voltage that is applied to the non-inverting

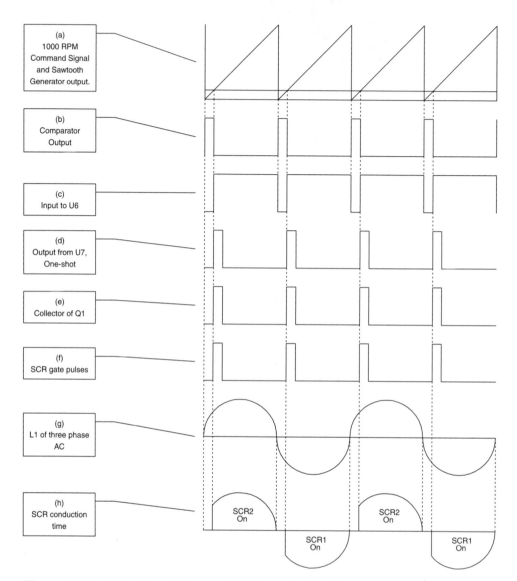

Figure 11-8: *Waveform timing and SCR conduction time at 1000 RPM.*

input of comparator U5. The DC voltage level shown is representative of a motor speed of 1000 RPM. Notice that this level is lower than that shown in Figure 11-7a. This is a result of the slower motor speed and lower output voltage from the tachometer-generator. These two inputs cause the output of the comparator to produce a pulsating DC as shown in Figure 11-8b. Notice that the pulse is initially positive, and is switched off when the two inputs of the comparator become equal. The output of U5 remains off until the sawtooth generator resets. At this point, the output of U5 switches on, and remains on until the sawtooth signal is equal to the feedback voltage. Compare the width of the positive pulses in Figure 11-8b with those in Figure 11-7b. Notice that the positive pulses are narrower in Figure 11-8b.

The pulsating output of U5 is fed through diode CR2 to the input of U6. Because of CR2, the input signal to U6 will be inverted compared to the output of U5. This is shown in Figure 11-8c. Again, compare the positive pulses in Figure 11-8c with those shown in Figure 11-7c. Notice that the positive pulses are wider in Figure 11-8c than those shown in Figure 11-7c. The positive pulses from the output of U6 trigger the one-shot, U7. The output of the one-shot, U7, will appear in phase with the input pulses as seen in Figure 11-8d. Notice that the output pulse of the one-shot now occurs earlier than it did in Figure 11-7d. This will cause the SCRs to fire sooner, causing them to conduct for a longer period of time. Let's see if this is what occurs.

The output of the one-shot, U7, is fed through diode CR3 to the base of the Darlington transistor, Q1. Notice in Figure 11-8e that the collector of Q1 is in phase with the output of the one-shot. When Q1 conducts, current will flow through the pulse transformer, T1.

When current flows through pulse transformer T1, a pulse appears at the gates of SCR1 and SCR2. This is shown in Figure 11-8f. Recall that only the SCR that is properly biased will conduct. The SCR that is triggered and properly biased will conduct until the AC voltage drops to the zero crossing point. The next pulse from T1 will cause the other SCR to conduct. Therefore, the SCRs will conduct alternately as shown in Figure 11-8g. Notice that the SCRs conduct for a longer amount of time as compared to the conduction of the SCRs in Figure 11-7g. This results in a higher average voltage applied to the AC motor. This will cause the speed of the AC motor to increase.

Let's assume that a decrease in load has caused the motor speed to increase to 2600 RPM. Refer to Figure 11-9. The waveform in Figure 11-9a shows the sawtooth waveform at the inverting input of U5, the comparator. At the same time, Figure 11-9a also shows the DC voltage that is applied to the non-inverting input of comparator U5. The DC voltage level shown is representative of a motor speed of 2600 RPM. Notice that this level is higher than that shown in Figure 11-7a. This is a result of the faster motor speed and higher output voltage from the tachometer-generator. These two inputs cause the output of the comparator to produce a pulsating DC as shown in Figure 11-9b. Notice that the pulse is initially positive, and is switched off when the two inputs of the comparator become equal. The output of U5 remains off until the sawtooth generator resets. At this point, the output of U5 switches on, and remains on until the

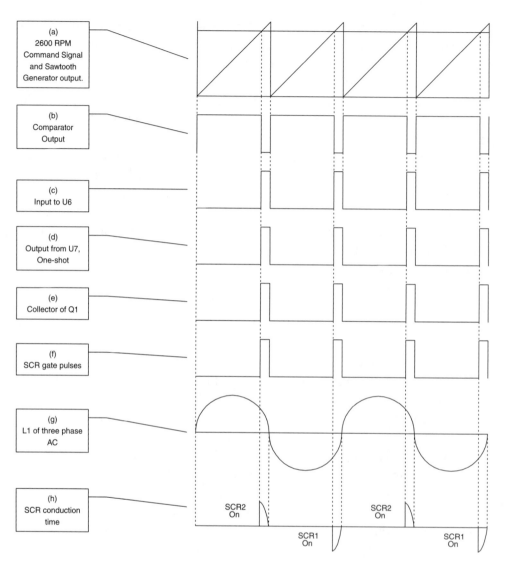

Figure 11-9: *Waveform timing and SCR conduction time at 2600 RPM.*

sawtooth signal is equal to the feedback voltage. Compare the width of the positive pulses in Figure 11-9b with those in Figure 11-7b. Notice that the positive pulses are wider in Figure 11-9b.

The pulsating output of U5 is fed through diode CR2 to the input of U6. Because of CR2, the input signal to U6 will be inverted compared to the output of U5. This is shown in Figure 11-9c. Again, compare the positive pulses in Figure 11-9c with those shown in Figure 11-7c. Notice that the positive pulses are narrower in Figure 11-9c than those shown in Figure 11-7c. The positive pulses from the output of U6 trigger the one-shot, U7. The output of the one-shot, U7, will appear in phase with the input pulses as seen in Figure 11-9d. Notice that the output pulse of the one-shot now occurs later than it did in Figure 11-7d. This will cause the SCRs to fire later, causing them to conduct for a shorter period of time. Let's see if this is what occurs.

The output of the one-shot, U7, is fed through diode CR3 to the base of the Darlington transistor, Q1. Notice in Figure 11-9e that the collector of Q1 is in phase with the output of the one-shot. When Q1 conducts, current will flow through the pulse transformer, T1.

When current flows through pulse transformer T1, a pulse appears at the gates of SCR1 and SCR2. This is shown in Figure 11-9f. Recall that only the SCR that is properly biased will conduct. The SCR that is triggered and properly biased will conduct until the AC voltage drops to the zero crossing point. The next pulse from T1 will cause the other SCR to conduct. Therefore, the SCRs will conduct alternately as shown in Figure 11-9g. Notice that the SCRs conduct for a shorter amount of time as compared to the conduction of the SCRs in Figure 11-7g. This results in a lower average voltage applied to the AC motor. This will cause the speed of the AC motor to decrease.

Review Questions

1. Does a pulse width modulated variable voltage inverter drive maintain a constant V/Hz ratio? Explain.

2. Describe the effect on motor speed of varying the frequency of the AC. Describe the effect on torque.

3. Describe the effect on motor speed of varying the stator voltage. Describe the effect on torque.

4. List several ways in which this PWM VVI drive is similar to the SCR armature voltage controller introduced in Chapter 3.

5. Explain how the firing angles of the SCRs are varied.

Chapter 12

Current Source Inverters

OBJECTIVES

After completing this chapter, you will be able to:

■ Discuss the differences between a VVI and a current source inverter.

■ Identify two basic methods of operation of a current source inverter.

■ Discuss several advantages and disadvantages of the current source inverter.

■ Discuss the theory of operation of a current source inverter.

In this chapter, we will discuss the current source inverter, also known as the current fed inverter. We will see how the current source inverter differs from the voltage source inverter studied in Chapters 10 and 11. We will also see how the current source inverter works. Finally, we will consider the advantages and disadvantages of the current source inverter drive.

What Is a Current Source Inverter?

The **current source inverter (CSI),** or **current fed inverter,** is a type of inverter drive in which the current is controlled while the voltage is varied to satisfy the motor's needs. To accomplish these tasks, a large inductor is used in the DC link section. Recall that in the variable voltage inverter a large capacitor was used to keep the DC voltage constant. A capacitor opposes a change in voltage. In the current source inverter a large inductor is used to keep the DC current constant. Remember that an inductor opposes a change in current.

How Does the CSI Work?

A CSI can operate in two basic ways. One is by means of a phase-controlled bridge rectifier circuit; the other is via a diode bridge rectifier circuit and chopper control. We will examine the phase-controlled bridge rectifier type shown in Figure 12-1 first.

From previous chapters you should recognize the SCRs used in the bridge rectifier circuit, which allows us to vary the firing angle of the SCRs. By varying the firing angle, we can cause the SCRs to conduct later in their cycle, thus producing a lower average output current to the DC bus. Conversely, if we fire the SCRs earlier in their cycle, we will produce a higher average output current to the DC bus. The variable DC current is fed to the large inductor in the DC bus section. This inductor provides a constant current to the inverter section. The inverter then provides either six-step control, or pulse width modulation control, as discussed in Chapter 11.

The other operating method used in a current source inverter is a diode bridge rectifier with chopper control. Figure 12-2 shows how this method works. Notice that a standard, run-of-the-mill diode bridge is used in the converter stage to rectify the AC into DC. The DC is then fed into a chopper circuit. Recall from Chapter 5 that a chopper is basically an electronic switch that turns on and off rapidly, "chopping" the DC. In this way the chopper circuit can vary the DC current. If the chopper is closed (or conducting) longer than it is open (not

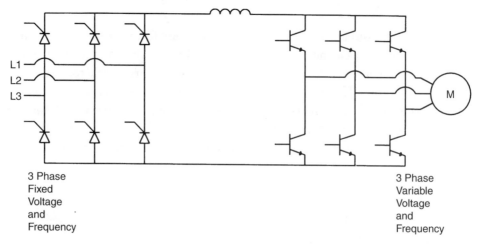

3 Phase
Fixed
Voltage
and
Frequency

3 Phase
Variable
Voltage
and
Frequency

Figure 12-1: *Current source inverter with a phase-controlled bridge rectifier.*

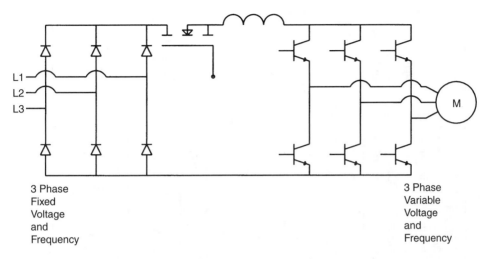

Figure 12-2: *Current source inverter with a diode bridge rectifier and chopper control.*

conducting), a higher average DC current flows. If the chopper is open (not conducting) longer than it is closed (or conducting), a lower average DC current flows. This variable DC current is fed to the large inductor in the DC bus section. The inductor provides the inverter stage with a constant current. As before, the inverter can be either a six-step inverter or a pulse width modulated inverter.

What Are the Advantages and Disadvantages of Current Source Inverters?

Current source inverters have several distinct advantages over variable voltage inverters. They provide protection against short circuits in the output stage; they can handle oversized motors. In addition, CSIs have relatively simple control circuits and good efficiency. As to their disadvantages, CSIs produce torque pulsations at low speed, cannot handle undersized motors, and are large and heavy. The phase-controlled bridge rectifier CSI is less noisy than its chopper controlled counterpart, does not need high-speed switching devices, and cannot operate from batteries. The chopper controlled CSI can operate from batteries, and produces more noise as a result of its need for high-speed switching devices. Refer to the chart in Figure 12-3 on the next page for a quick comparison between the various types of variable voltage and current source inverters.

Features	Variable Voltage Inverter with Phase Control	Variable Voltage Inverter with Chopper Control	Current Source Inverter with Phase Control	Current Source Inverter with Chopper Control	Pulse Width Modulation
Open Circuit Protection	YES	YES			YES
Short Circuit Protection			YES	YES	
Ability to Handle Oversized Motors			YES	YES	
Ability to Handle Undersized Motors	YES	YES			YES
Multiple Motor Applications	YES	YES			YES
Low-Speed Torque Pulsations	YES	YES	YES	YES	
Requires High-Speed Switching Devices		YES		YES	YES
Battery Operation		YES		YES	YES
Regenerative Operation	YES		YES		
Low-Speed Efficiency	GOOD	GOOD	GOOD	GOOD	MEDIUM
Complex Control Circuit	MEDIUM	MEDIUM	MEDIUM	MEDIUM	HIGH
Size and Weight	MEDIUM	MEDIUM	HIGH	HIGH	LOW

Figure 12-3: Inverter comparison chart.

Review Questions

1. What do the letters CSI stand for?

2. What is another name for a CSI drive?

3. Describe the device responsible for maintaining a constant current to the inverter section and explain how it does so.

4. List several differences between a CSI and a VVI.

5. Name two different methods of producing a variable current to the DC bus and explain how they work.

6. The inverter(s) that does/do not produce torque pulsations at low speeds is/are the

   a. VVI with phase control.

   b. VVI with chopper control.

   c. CSI with phase control.

   d. CSI with chopper control.

   e. pulse width modulation.

7. The inverter(s) that can operate from batteries is/are the

   a. VVI with phase control.

   b. VVI with chopper control.

   c. CSI with phase control.

   d. CSI with chopper control.

   e. pulse width modulation.

8. The inverter(s) that provide(s) short circuit protection is/are the

   a. VVI with phase control.

   b. VVI with chopper control.

   c. CSI with phase control.

   d. CSI with chopper control.

   e. pulse width modulation.

9. The inverter(s) that does/do not need high-speed switching devices is/are the

   a. VVI with phase control.

   b. VVI with chopper control.

   c. CSI with phase control.

   d. CSI with chopper control.

   e. pulse width modulation.

10. The inverter(s) that is/are not as efficient at low speeds is/are the

   a. VVI with phase control.

   b. VVI with chopper control.

   c. CSI with phase control.

   d. CSI with chopper control.

   e. pulse width modulation.

11. The inverter(s) that use(s) more complex control circuitry is/are the

   a. VVI with phase control.

   b. VVI with chopper control.

   c. CSI with phase control.

   d. CSI with chopper control.

   e. pulse width modulation.

12. The inverter(s) that can handle oversized motors is/are the

   a. VVI with phase control.

   b. VVI with chopper control.

   c. CSI with phase control.

   d. CSI with chopper control.

   e. pulse width modulation.

13. The inverter(s) that does/do not produce high-frequency noise is/are the

   a. VVI with phase control.

   b. VVI with chopper control.

   c. CSI with phase control.

   d. CSI with chopper control.

   e. pulse width modulation.

14. The biggest and heaviest inverter(s) is/are the

    a. VVI with phase control.

    b. VVI with chopper control.

    c. CSI with phase control.

    d. CSI with chopper control.

    e. pulse width modulation.

Chapter 13

Flux Vector Drives

OBJECTIVES

After completing this chapter, you will be able to:

- Discuss, in general, the theory of current vectors in an AC induction motor.

- Discuss, in general, the theory of operation of a flux vector drive.

- Identify some benefits and disadvantages of a flux vector drive.

Because this text was written with the maintenance technician in mind care has been taken to approach the theory behind the operation of electronic variable speed drives in a simple, straightforward manner. However, because one rather complicated type of drive, the flux vector drive, is gaining in popularity, this chapter discusses its operation. Figure 13-1 shows a SECO model VR flux vector drive. Figure 13-2 shows the control section of a SECO VR flux vector drive. Figure 13-3 shows the power section of a SECO VR flux vector drive. Due to the complexity of its nature, the theory behind this drive is somewhat beyond the scope of this text. However, for the individual exposed to flux vector drives, this section provides a basic understanding of their operation. We will begin our discussion of flux vector drives with a slightly different look at general motor theory.

To explain the operation of a typical, three-phase, AC induction motor, we always speak in terms of the rotating magnetic field. In order to understand flux vector drives, we must delve a little deeper into the characteristics of the rotating magnetic field.

Figure 13-4 on page 127 shows a typical three-phase sine wave. Notice also the vectors drawn to represent the balanced three-phase currents. (Recall that a vector represents both magnitude and direction.) Observe that these vectors are drawn 120° apart to represent the normal phase shift in a three-phase system. Note, too, that at 30° vectors L1 and L2 are drawn with their arrowheads pointing outward, away from the center, or neutral point. Vector L3 is drawn with its arrowhead pointing inward, or toward the neutral point. The direction

Figure 13-1:   *SECO model VR flux vector drive.*

of these arrowheads corresponds to the polarity of the instantaneous current. If the current is in the positive portion of the sine wave, the arrowhead is drawn pointing outward or away from the neutral. If the current is in the negative portion of the sine wave, the arrowhead will be drawn pointing inward or toward the neutral point. Also, the length of the vectors varies to represent the magnitude of the current.

Now look at Figure 13-5 on page 127, which includes another set of drawings that show the addition of the current vectors and the resulting magnetic flux. (Recall that vectors are added by positioning them head to tail.) Notice that, as

Figure 13-2: *Control section of a SECO VR flux vector drive.*

we advance from 30° to 90°, the current vectors appear to rotate in a counter-clockwise direction, causing the flux to rotate in the same direction. As we continue to advance from 90° to 150°, the current vectors and the flux continue to rotate counterclockwise. If we continue to advance to 360°, the current vectors and the flux will complete one full revolution. All of this occurs in a predictable manner because the three-phase supply is balanced and sinusoidal and it has a constant amplitude and a constant frequency. What happens when we introduce some changes to this configuration?

Figure 13-3: *Power section of a SECO VR flux vector drive.*

Before we discuss changes to the frequency, amplitude, or phase rotation, be aware that the AC sine wave supplied to the motor is artificially produced by an inverter stage in the flux vector drive. This inverter stage thus allows us to exercise control over the synthetic AC and to make the changes described in the remainder of this chapter.

In Figure 13-6, we have caused the devices that produce the artificial AC to "jump ahead." In other words, we have advanced the AC from 60° to 90° instantaneously. This change has no effect on the direction of rotation or on the

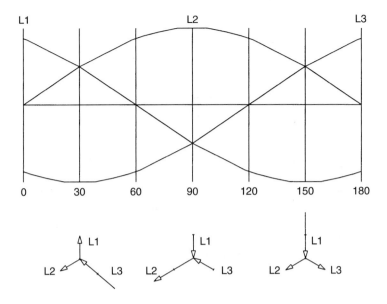

Figure 13-4:   *Typical three-phase AC with current vectors.*

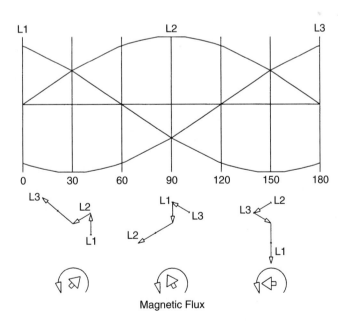

Magnetic Flux

Figure 13-5:   *Typical three-phase AC showing addition of the current vectors and the resulting magnetic flux.*

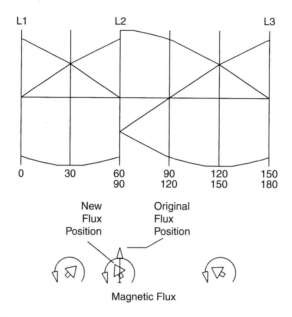

Figure 13-6: *Change in magnetic flux position due to a step change.*

strength of the magnetic flux. This does, however, cause the motor suddenly to advance 30° in rotation. Because we can control the firing sequence of the devices that produce the synthetic AC, we can also control the amount of advancement of the flux field. This jump is known as a *step change*.

Given this control over the devices that produce the synthetic AC, what happens when we hold these same devices on for an extended period of time? Referring to Figure 13-7, notice that at the 60° point the step change discussed previously does not occur, nor have we allowed the current to advance normally. In this instance, we have held the current constant for 30°. As a result, the flux rotation has been stopped or held stationary for the 30° period.

You should also recognize that by taking advantage of our control over the AC, we can change the phase rotation and thus the direction of rotation of the motor as well.

Flux vector drives operate by monitoring rotor position and changing the rotor's position as compared to its orientation to the stator field. To know the rotor's position, encoders are typically used as feedback devices. In addition, current feedback from two of the three phases is used. This amount and variety of feedback is necessary to achieve the variety of control mechanisms that a flux vector drive offers. For instance, if we vary the amplitude of the current, we will vary the magnitude of the current vectors. If we vary the phase rotation, we will vary the direction of rotation. Causing a step change to occur will result in an

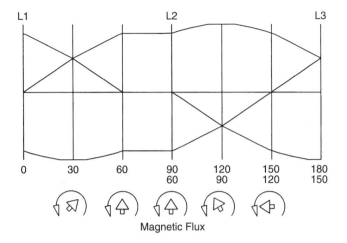

Figure 13-7: *Magnetic flux held constant for 30°.*

instantaneous advance in motor position. Finally, if we hold the phase current constant, we will cause the motor to hold a constant position.

The flux vector drive must constantly monitor the motor's performance and allow for high speed corrections to occur to maintain accuracy in motor positioning and performance. For this reason, flux vector drives are matched to the particular motor that they will operate. Although this makes flux vector drives more expensive, the resulting improved performance in the areas of rapid response and precise control makes these drives well worth their price.

Review Questions

1. A vector represents both _____ and _____.

2. What is the term used to describe the "jumping ahead" of the artificial AC in a flux vector drive?

3. What happens when a "jump ahead" does not occur, nor is a change allowed in the flux field?

4. **True or False?** Flux vector drives must be matched to the particular motor that they operate.

5. **True or False?** Flux vector drives are no more expensive than their counterparts.

6. **True or False?** Flux vector drives provide rapid response and precise control.

Chapter 14

Troubleshooting
Inverter Drives

OBJECTIVES

After completing this chapter, you will be able to:

■ List the four main areas to check when troubleshooting an AC inverter drive system.

■ Discuss several safety considerations when troubleshooting an AC inverter drive system.

■ Define the term phase imbalance.

■ Calculate phase imbalance.

■ List several possible causes of AC inverter drive system failure.

■ List several types of adjustments that can be performed on AC inverter drives.

■ Explain the purpose of various adjustments that can be performed on AC inverter drives.

In this chapter, we will consider troubleshooting AC inverter drives, identify the four main areas of possible problems, and present some techniques for accurately and rapidly fixing these problems.

Even though some of the techniques presented in this chapter were covered in Chapter 8, we will repeat them here as a review. In particular, those concepts regarding safety and common sense cannot be emphasized enough.

What Are the Four Main Areas of Possible Problems?

In an AC inverter drive the four main areas of possible problems are the same as those in a DC drive:

1. the electrical supply to the motor and the drive.

2. the motor and/or its load.

3. the feedback device and/or sensors that provide signals to the drive.

4. the drive itself.

Even though the problem may be in any one or several of these areas, the best place to begin troubleshooting is the drive unit itself. The reason for this is that most drives have some type of display that aids in troubleshooting. This display may be simply an LED that illuminates to indicate a specific fault condition, or it may be an error or fault code that can be looked up in the operator's manual. For this reason, *it is strongly suggested* that a copy of the fault codes be made and fastened to the inside of the drive cabinet, where it will be readily available to the maintenance technician. The original should be placed in a safe location such as the maintenance supervisor's office.

Before you proceed, *stop and think about what you are doing!* Before working on any electrical circuit, *remove all power.* Sometimes removing the power is not possible or permissible. In these instances, *work carefully and wear the appropriate safety equipment.* Do not rely on safety interlocks, fuses, or circuit breakers to provide personnel protection. *Always use a voltmeter to verify that the equipment is de-energized, and tag and lock the circuit out!*

Even when the power has been removed, you are still subject to shock and burn hazards. Most drives have high-power resistors inside, and these resistors can and do get *hot!* Give them time to cool down before touching them. Most drives also have large electrolytic capacitors. These capacitors can and do store an electrical charge. Usually the capacitors have a bleeder circuit that dissipates this charge. However, this circuit may have failed. *Always verify that electrolytic capacitors are fully discharged.* Do this by measuring carefully any voltage present across the capacitor terminals. If voltage is present, *use an approved shorting device to discharge the capacitor completely.*

Now slow down, take your time, and use your senses, even though production has stopped, and you are under pressure to fix the equipment. Sometimes a little extra initial time to take stock of the situation can save considerable time later. So do the following:

Look! Do you see any charred or blackened components? Have you noticed any arcing? Do fuses or circuit breakers appear blown or tripped? Do you see any discoloration around wires, terminals, or components? A good visual inspection can save a lot of troubleshooting time.

Listen! Did you hear any funny or unusual noises? A frying or buzzing sound may indicate arcing. A hum may be normal or an indication of loose laminations in a transformer core. A rubbing or chafing sound may indicate that a cooling fan is not rotating freely.

Smell! Do you notice any unusual odors? Burnt components and wires give off a distinctive odor. Metal will smell hot if subjected to too much friction.

Touch! (Be very careful with this!) If components feel cool, that may indicate that no current is flowing through the device. If components feel warm, chances are that everything is normal. If components feel hot, things may be normal too, although more than likely too much current is flowing, or too little cooling is taking place. In any event, there may be a problem worth further investigation.

The point of all this is that by being *observant* you have a good chance of discovering the problem or problem area. Before we consider in more detail the four main areas mentioned earlier, review the proper techniques for installing an inverter.

First you should make the AC power connections. Be careful to use the proper voltage and connect the power source to the correct terminals on the inverter. If the supply voltage is different from the voltage indicated on the inverter, you may have to move some jumpers on the inverter to reconfigure it for the proper voltage levels.

After the AC supply has been connected, make the motor connections. Again, be careful to connect the motor leads to the proper terminals on the inverter. You should also verify that the voltage rating of the motor matches the voltage rating of the inverter.

Next you must program or otherwise set the proper operating parameters for the inverter. We will now examine how to accomplish this in more detail.

Depending on the model, you may need to set the inverter for remote control (start/stop/adjust speed from a remote location) or local control (start/stop/adjust speed from the inverter's control panel).

If you are using remote control, you will need to wire the control circuit. You will also need to wire the speed/torque control potentiometer.

The preceding instructions are general guidelines for installing an inverter. Check the manual that comes with the particular inverter that you are using and follow the procedures presented there precisely.

Now let's examine some of the parameters that can be programmed into an inverter. Keep in mind that not all inverters share these parameters. Some manufacturers list the same parameters in different order. Certain manufacturers have other names for the settings discussed here. Also, keep in mind that some inverters will have more parameters and others will have fewer parameters than those mentioned here. Again, familiarize yourself with the particular inverter that you are using. In general, programmable inverter parameters include the following.

Analog input select: Sets either a 0–10 V or 4–20 mA reference that is proportional to the frequency input.

Analog output select: Provides a 4–20 mA output signal that is proportional to the motor speed.

Basic set-up: Can be set for constant torque or constant speed.

Current: The output current of the inverter.

Current limit: Sets the maximum current available to the motor. If the setting is at the maximum permissible value, the motor will have maximum starting torque. This value can be on the order of 160% of nominal motor current. If the current limit is set too low, the inverter can trip out.

DC brake time: If selected, this parameter will provide additional braking torque at low motor speeds.

Digital input select: If selected, this value bypasses the ramp time down setting, and the inverter decelerates the motor in the shortest possible amount of time as a result.

Frequency: The output frequency of the inverter.

Jogging speed: Sets the speed of the motor when jogging.

Local/remote: Can be set for local control from the inverter's control panel or remote control from a start/stop station located away from the inverter cabinet.

Maximum speed: Depending on the setting, it may be possible to attain a speed higher than the rated speed of the motor. This parameter must be set higher than the minimum speed setting. If it is set lower than the minimum speed, the motor will not run.

Minimum speed: Slowest speed setting at which the motor will run.

Motor magnetization: Set to the no load current rating on the motor's nameplate.

Motor nominal current: Set to the full load current rating on the motor's nameplate.

Motor nominal frequency: The rated frequency of the motor from the motor nameplate. This value should be set as closely as possible to the specified value.

Motor nominal voltage: The rated line voltage of the motor from the motor nameplate. This value should be set as closely as possible to the specified value.

Motor power: This is the motor's power rating expressed either in horsepower (HP) or in kilowatts (kw). Some drives will accept the value in either unit of measurement while others require converting the units from one measure to the other.

Ramp time down: This is the deceleration time (the time required to get from maximum speed to minimum speed) expressed in seconds. If this time is set too short, the inverter can trip out.

Ramp time up: This is the acceleration time (the time required to get from minimum speed to maximum speed) expressed in seconds. If this time is too short, the inverter can trip out.

Relay output select: Provides a contact closure when the inverter is placed in the run mode.

Slip compensation: Typically the factory setting for this value should be adequate. This setting is affected by the motor power, motor nominal voltage, and motor nominal frequency values.

Start compensation: Typically the factory setting for this value should be adequate. This setting is affected by the motor power, motor nominal voltage, and motor nominal frequency values.

Start voltage: Typically, the factory setting for this value should be adequate. This setting is affected by the motor power, motor nominal voltage, and motor nominal frequency values.

Start/stop mode: This value is programmed for the various types of start/stop circuits used. For example, a two-wire start/stop, a three-wire start/stop, a three-wire start/stop with a jog, and so on.

Thermal motor protect: Depending on the setting chosen, this parameter will either flash the display when the motor's critical temperature is reached or trip the inverter.

Torque: Calculated motor torque. Dependent on the programmed settings of the motor nominal current and the motor magnetization.

Trip reset mode: If selected, this parameter will prevent the inverter from restarting automatically after a trip.

V/f ratio: Typically the factory setting for this value should be adequate. This setting is affected by the motor power, motor nominal voltage, and motor nominal frequency values.

Voltage: The output voltage of the inverter.

Where Should Troubleshooting of an Inverter Drive System Begin?

In order to begin troubleshooting, we need to understand the general, step-by-step sequence of events that most inverter drive systems follow on start-up and shutdown.

We begin with the AC power applied to the inverter. We must adjust the reference or set point for the desired speed and torque characteristics of the motor. Next the start/stop circuit is placed in the start or run mode. Instantly, the main control components begin a diagnostic routine. We will examine some of the routine's fault codes shortly. If no faults are detected, the driver activates the power semiconductors. These produce the output frequency and V/f ratio programmed into the inverter to match the speed and torque settings. The programmed setting for the ramp time up will control how long it takes the motor to reach the desired speed. While the motor speed is ramping up, the motor current is monitored. Should the current exceed the programmed current limit setting, the inverter may (depending on the manufacturer), automatically adjust the ramp time up program or simply trip. If tripping occurs repeatedly, the ramp time up program may need to be modified to accommodate a longer acceleration time. The ramp time up setting will cause the inverter output voltage and frequency to increase, accelerating the motor to the desired speed set point.

When it becomes necessary to stop the motor, most inverters offer several options. One option is basically to let the motor coast to a stop. This can take a very long time for high-inertia loads. The ramp down time setting allows the inverter to slow the motor gradually. This deceleration is accomplished by allowing the motor to feed its self-generated energy back into the inverter. The inverter will use large resistors to absorb this energy. This process, called dynamic braking, is usually insufficient to bring the motor to a controlled and rapid stop because as the motor slows, less energy is self-generated. Therefore it is common practice to use a mechanical brake in conjunction with dynamic braking to provide additional braking action at slow speeds. Another method of stopping the motor without using a mechanical brake is **plugging** or **DC injection.** When the operator wishes to stop the motor, DC current is fed into the motor winding. This current replaces the rotating magnetic field with a fixed magnetic field. The rotor soon becomes locked with this fixed magnetic field, which in effect stops the motor rotation. This method should not be used repeatedly because heat can build up in the motor as a result and damage the windings. Now that we have a basic understanding of the processes that an inverter follows on start-up and shutdown, let's investigate the types of problems that can be encountered in the four main areas of an inverter drive system.

The Electrical Supply to the Motor and the Drive

Most maintenance technicians believe that the power distribution in an industrial environment is reliable, stable, and free of interference. Nothing could be further from reality! Frequent outages, voltage spikes and sags, and electrical noise are normal operating occurrences. The effect of these is not as detrimental to motor performance as it can be to the operation of the drive itself. Most AC inverters are designed to operate despite variations in supply voltages. Typically the incoming power can vary as much as ±10% with no noticeable change in drive performance. However, in the real world it is not unusual for power line fluctuations to exceed 10%. These fluctuations may occasionally cause a controller to trip. If tripping occurs repeatedly, a power line regulator may be required to hold the power at a constant level.

A power line regulator will be of little use, however, should the power supply to the controller fail. In this situation a UPS (Uninterruptible Power Supply) is needed. Several manufacturers produce a complete power line conditioning unit. These units combine a UPS with a power line regulator.

Quite often controllers are connected to an inappropriate supply voltage. For example, it is not unusual for a drive rated 208 V to be connected to a 240 V supply. Likewise, a 440 V rated drive may be connected to a 460 V or even a 480 V source. Usually the source voltage should not exceed the voltage rating of the drive by more than 10%. For a drive rated at 208 V the maximum supply voltage is 229 V (208 × 10% = 20.8 + 208 = 229). Obviously the 208 V drive, when connected to the 240 V supply, is receiving excess voltage and should not be used. For our 440 V drive the maximum supply voltage is 484 V (440 × 10% = 44 + 440 = 484). Although this value appears to fall within permissible limits, another potential problem exists here. Suppose that the power line voltage fluctuates by 10%. If the 440 V source suffers a 10% spike, the voltage will increase to 484 V. This value is within the design limits of the drive. But what happens when we connect the 440 V drive to a 460 V or a 480 V power line? If we experience that same 10% spike, the 460 V line will increase to 506 V (460 × 10% = 46 + 460 = 506), and the 480 V line will increase to 528 V (480 × 10% = 48 + 480 = 528)! We have thus exceeded the voltage rating of the drive and probably damaged some internal components! Most susceptible to excess voltage and spikes or transients are the SCRs, MOSFETs, and power transistors. Premature failure of capacitors can also occur. As you can see, it is very important to match the line voltage to the voltage rating of the drive.

An equally serious problem occurs when the phase voltages are unbalanced. Typically, during construction care is taken to balance the electrical loads on the individual phases. As time goes by and new construction and remodeling occurs, it is not unusual for the loading to become imbalanced, causing intermittent tripping of the controller and perhaps premature failure of components. To determine if an imbalanced phase condition exists you will need to do the following:

1. Measure and record the *phase voltages* (L1 to L2, L2 to L3, and L1 to L3).

2. Add the three voltage measurements from step 1 and record the *sum of all phase voltages.*

3. Divide the sum from step 2 by 3 and record the *average phase voltage.*

4. Now subtract the average phase voltage obtained in step 3 from each phase voltage measurement in step 1 and record the results. (Treat any negative answers as positive answers.) These values are the *individual phase imbalances.*

5. Add the individual phase imbalances from step 4 and record the *total phase imbalance.*

6. Divide the total phase imbalance from step 5 by 2 and record the *adjusted total phase imbalance.*

7. Now divide the adjusted total phase imbalance from step 6 by the average phase voltage from step 3 and record the *calculated phase imbalance.*

8. Finally, multiply the calculated phase imbalance from step 7 by 100 and record the *percent of total phase imbalance.*

Consider an example involving a 440 V three-phase supply to an AC inverter drive to see how this process works.

1. L1 to L2 = 432 V; L2 to L3 = 435 V; and L1 to L3 = 440 V.

2. The sum of all phase voltages equals 432 V + 435 V + 440 V, or 1307 V.

3. The average phase voltage is equal to 1307 V ÷ 3, or 435.7 V.

4. To find the individual phase imbalances, we subtract the average phase voltage from the individual phase voltages and treat any negative values as positive. So L1 to L2 = 432 V − 435.7 V, or 3.7 V; L2 to L3 = 435 V − 435.7 V, or 0.7 V; and L1 to L3 = 440 V − 435.7 V, or 4.3 V.

5. Now we find the total phase imbalances by adding the individual phase imbalances: 3.7 V + 0.7 V + 4.3 V = 8.7 V.

6. To find the adjusted total phase imbalance we divide the total phase imbalance by 2: 8.7 V ÷ 2 = 4.35 V.

7. Next we find the calculated phase imbalance by dividing the adjusted total phase imbalance by the average phase voltage: 4.35 V ÷ 435.7 V = 0.00998.

8. Finally we multiply the calculated phase imbalance by 100 to find the percent total phase imbalance: 0.00998 × 100 = 0.998%.

In this example we are within tolerances, and the differences in the phase voltages should not cause any problems. In fact, as long as the percent total phase imbalance does not exceed two percent, we should not experience any difficulties as a result of the differences in phase voltages.

What Problems Can Occur with the Motor and the Load?

Probably the most common cause of motor failure is heat! Heat can occur simply as a result of the operating environment of a motor. Many motors are operated in areas of high ambient temperature. If steps are not taken to keep the motor within its operating temperature limits, the motor will fail. Some motors have an internal fan that cools the motor. If the motor is operated at reduced speed, this internal fan may not turn fast enough to cool the motor sufficiently. In these instances, an additional external fan may be needed to provide additional cooling to the motor. Typically, such fans are interlocked with the motor operation in such a way that the motor will not operate unless the fan operates as well. Therefore a fault in the external fan control may prevent the motor from operating.

The sensors used to sense motor temperatures consist simply of a nonadjustable thermostatic switch that is normally closed and opens when the temperature rises to a certain level. In this event you must wait for the motor to cool down sufficiently before resetting the temperature sensor and restarting the motor.

Periodic inspection of the motor and any external cooling fans is strongly recommended. The fans should be checked for missing or bent vanes. All openings for cooling should be kept free of obstructions. Any accumulation of dirt, grease, or oil should be removed. If filters are used, these must be cleaned or replaced on a routine schedule.

Heat may also cause other problems. When motor windings become overheated, the insulation on the wires may break down. This breakdown may cause a short, which may lead to an open condition. A common practice used to find shorts or opens in motor windings is to megger the windings with a megohmmeter. Extreme caution must be taken when using a megger on the motor leads. Be certain that you have disconnected the motor leads from the drive. Failure to do this will cause the megger to apply a high voltage to the output section of the drive. Damage to the power semiconductors will result. You may also decide to megger the motor leads at the drive cabinet. Again, be certain that you have disconnected the leads from the drive unit, and megger only the motor leads and motor winding. *Never megger the output of the drive itself!*

Since the motor drives a load of some kind, it is also possible for the load to create problems. The drive may trip out if the load causes the motor to draw an excessive amount of current for too long a time. When this occurs, most drives display some type of fault indication. The problem may be a result of the motor operating at too high a speed. Quite often, a minor reduction in speed is all that is necessary to prevent repeated tripping of the drive. The same effect occurs if the motor is truly overloaded. Obviously, in this case either the motor size needs to be increased or the size of the load decreased to prevent the drive from tripping.

Some loads have a high inertia. They require not only a large amount of energy to move, but once moving, a large amount of energy to stop. If the drive cannot provide sufficient braking action to match the inertia of the load, the drive may trip, or overhaul. A drive with greater braking capacity is needed in such cases to prevent tripping from recurring.

Can Feedback Devices or Sensors Cause Problems?

Mechanical vibration may cause the mounting of feedback devices to loosen and their alignment to vary. Periodic inspections are necessary to verify that these devices are aligned and mounted properly.

It is also important to verify that the wiring to these devices is in good condition and the terminations are clean and tight. Another consideration regarding the wiring of feedback devices is electrical interference. Feedback devices produce low-voltage/low-current signals that are applied to the drive. If the signal wires from these devices are routed next to high-power cables, interference can occur. This interference may result in improper drive operation. To eliminate the possibility that this will occur, several steps must be taken. The signal wires from the feedback device should be installed in their own conduits. Do not install power wiring and signal wiring in the same conduit. The signal wires should be shielded cable, with the shield wire grounded to a good ground at the drive cabinet only. Do not ground both ends of a shielded cable. When the shielded signal cable is routed to its terminals in the drive cabinet, the cable should not be run or bundled parallel to any power cables, but instead at right angles to such power cables. Furthermore, the signal cable should not be routed near any high-power contactors or relays. When the coils of a contactor or relay are energized and de-energized, a spike is produced. This spike can create interference with the drive. To suppress this spike, it may be necessary to install a free-wheeling diode across any DC coils or a snubber circuit across any AC coils.

What Problems Can Occur in the Inverter Drive Itself?

First, look for fault codes or fault indicators. Most drives provide some form of diagnostics, and this can be a great timesaver.

Let's look at some of the fault codes, symptoms, probable causes, and fixes for some common problems:

The inverter is inoperable, and no LED indicators are illuminated. Possibly no incoming power is present. You can verify this by measuring the voltage at the power supply input terminals in the inverter cabinet. The problem may be caused by a blown fuse, an open switch, an open circuit breaker, or an open disconnect. There are several things to check if the fuses are blown. One item to look for is a shorted **metal oxide varistor (MOV),** a device that provides surge protection to the inverter. If a significant power surge occurred, the MOV may have shorted to protect the inverter. Because the MOV is located across the power supply lines (to provide protection), a shorted MOV can cause fuses to blow. Another reason why fuses blow is a shorted diode in the rectifier circuit. A shorted or leaky filter capacitor in the power supply may also cause fuses to blow.

The inverter is powered up but does not work. There are indications of a fault condition. The "watchdog" circuit may have tripped. Remember that a watchdog circuit monitors the power lines for disturbances, and if the disturbance occurs for a long duration, the watchdog circuit trips the controller to protect it from damage. A heavy starting load may sag the power line voltage to such a point that the inverter receives insufficient voltage. This deficiency can trip the inverter. Likewise, if the load has high inertia, it is possible for the re-generative effect to provide excess voltage to the inverter. Another way that the inverter may receive excess voltage is use of the power factor correction capacitors while the load is removed.

If the watchdog circuit has not tripped, other possible reasons exist for this fault condition. It is possible that there are interlocks on the cabinet or cooling fans, and one or more of these may be open. Likewise, a temperature sensor on a heatsink on the power semiconductors in the inverter or in the motor itself may have detected an excessive temperature condition and opened as a result.

The inverter is energized, and a fault is indicated. The motor does not respond to any control signals. If the load is too high for the motor settings, the motor may fail to rotate. It may be necessary to increase the current limit or voltage boost settings to allow the motor to overcome the load. Another possibility is that the load is overhauling the motor. If this is the situation, it will be necessary to adjust the deceleration time to allow the motor to take longer to brake the load. It may also be necessary to add auxiliary braking in the form of a mechanical brake. If the motor leads have developed a short or the motor itself has a shorted winding or is overloaded or stalled, the current limit sensor may trip the inverter. Tripping may also occur as a result of a shorted power semiconductor.

As you can see, you need to be aware of many areas when dealing with an inoperable inverter. Fortunately the inverter itself can help a great deal by displaying fault codes. The operator's manual will interpret the fault codes and give instructions for clearing the fault condition. Let's examine some fault codes and how they can help in your troubleshooting. Remember that not all manufacturers provide the same fault codes. Some provide more, and others provide fewer.

As mentioned earlier, one fault that may be displayed indicates an over current limit condition. This indicator should direct you to examine the motor for mechanical binding, jams, and so forth. To verify whether one of these conditions is the cause of the problem, disconnect the motor from the load and reset the inverter. If the fault clears, then you know that the load is the cause. If the fault reappears, you need to look further, perhaps at the motor itself.

Another fault code, over voltage, may be the result of a high-inertia load that causes overhauling. This fault code may also be a result of setting the deceleration ramp down parameter for rapid deceleration. Lengthening the deceleration time may clear the fault. If lengthening the time is not possible, additional mechanical braking may be required to bring the load to a rapid stop.

The inverter overload fault code is an indication of electrical problems. Examples of these are shorted or grounded motor leads or windings and/or defective power semiconductors. If the motor is suspect, disconnect it from the inverter. If the fault clears when you reset the inverter, you can assume that the problem lies in the motor and/or its leads. To verify whether the problem is in the power semiconductors, disconnect the gate lead from one of the devices. Reset the inverter. If the fault clears, you have found the problem. If the fault is still present, reconnect the gate lead, move to the next device, and disconnect its gate lead. Reset the inverter. Again, if the fault clears, you have found the problem. If the fault is still present, repeat the preceding steps until you have tested all of the power semiconductors.

Another fault code indicates shorted control wiring. If this code is displayed, simply disconnect the control wiring and reset the inverter. The fault should clear. This result indicates problems in the control wiring. If the fault does not clear, try unplugging the control board and resetting the inverter. A cleared fault condition in this case indicates problems in the control board.

It is a good idea to maintain an inventory of spare pc boards for the various inverters at your plant. That way, if you determine that the problem is caused by a defective pc board, it should be a fairly simple matter to replace the board with a spare. This will minimize down-time and allow you to try to repair the pc board in the shop, under a lot less pressure. If the pc board cannot be repaired, it may be possible to return it to the manufacturer for repair or exchange.

Although fault codes are excellent troubleshooting tools and can save a great deal of time, you should be aware of other potential problems that may or may not show up as fault codes, depending on the manufacturer.

Heat can produce problems in the drive unit. The cabinet may have one or more cooling fans, with or without filters. These fans are often interlocked with the drive power in such a way that the fan must operate in order for the drive to operate. Make certain that the fans are operational and the filters are cleaned or replaced regularly. The power semiconductors are typically mounted on heatsinks. A heatsink may have a small thermostat mounted on it to detect an excess temperature condition in the power semiconductors. If the heatsink becomes too hot, the thermostat will open, and the drive will trip. Usually these thermostats are self-resetting. You must wait for them to cool down and reset themselves before the drive will operate. If tripping occurs repeatedly, a more serious problem exists that requires further investigation.

If the drive is newly installed, problems are often the result of improper adjustments to the drive. On the other hand, if the drive has been in operation for some time, it is unlikely that readjustments are needed. All too often, an untrained individual will try to adjust a setting to "see if this fixes it!" Usually such adjustments only make things worse! This is not to say that adjustments are *never* needed. For example, if the process being controlled is changed or some component of the equipment has been replaced, it probably will be necessary to change the drive settings. For this reason it is very important to record the initial settings and any changes made over the years. This record should be placed in a safe location and a copy made and placed in the drive cabinet for easy access by maintenance personnel.

Replacing the drive should be the final choice. More often than not, if the drive is defective, something external to the drive is the reason for the drive's failure. Replacing the drive without determining what caused the failure may result in damage to the replacement drive. If the drive has failed, the possibility still exists that you can get it to work again.

Most drive failures occur in the power section, where you will find the power SCRs, transistors, MOSFETs, and so on. In some drives, these devices will be individual components. You can test these devices with reasonable accuracy using nothing more than an ohmmeter. (See Appendix C.) In other drives, the SCRs, transistors, and MOSFETs are contained within a power module.

An SCR power module is shown in Figure 14-1. One of these modules would be used for each phase of the three-phase AC. Notice that this module contains two SCRs. We will call the SCR to the left "A" and the SCR to the right "B." You can use an ohmmeter to test an SCR module with reasonable accuracy. To do so, you must understand what each terminal of the module represents. A schematic diagram of the SCR module is shown in Figure 14-2.

Notice in Figure 14-1 that there are a total of five terminals on the SCR module. Referring to Figure 14-2, you will see the same five terminals. Two of the terminals, (the small terminals in Figure 14-1), represent the gate connections for each SCR. The gate terminal toward the back of the module is connected to the gate of SCR "B," while the gate terminal at the front of the module is connected to the gate of SCR "A." The large terminal closest to the gate terminals

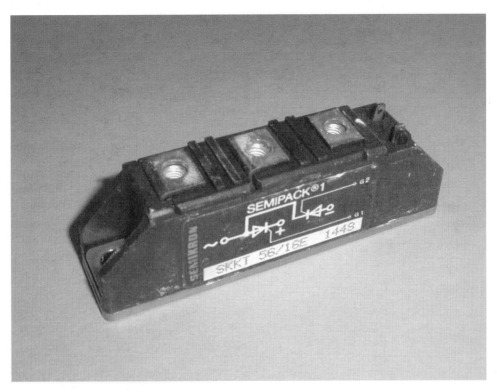

Figure 14-1: SCR power module.

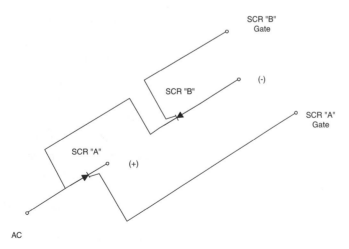

Figure 14-2: SCR power module schematic.

is the negative terminal of the module, which is also the anode of SCR "B." The large terminal farthest from the gate terminals is the AC terminal for the module, which is also the anode of SCR "A." Notice that the anode of SCR "A" is internally connected to the cathode of SCR "B." The large terminal in the center is the positive terminal of the module, which is also the cathode of SCR "A."

To test the module with an ohmmeter, follow these steps:

1. Place the ohmmeter in the "Ohms" position, if using a digital multimeters. (Use the 200 Ω position if using an analog multimeters.)

2. Place the black (negative) lead of your ohmmeter on the anode (terminal farthest from the gate terminals) of SCR "A."

3. Place the red (positive) lead of your ohmmeter on the cathode (center terminal) of SCR "A."

4. Your meter should indicate a very high or infinite resistance. A low resistance reading indicates a faulty SCR. Replace the module.

5. Reverse your meter connections. Place the black (negative) lead of your ohmmeter on the cathode (center terminal) of SCR "A."

6. Place the red (positive) lead of your ohmmeter on the anode (terminal farthest from the gate terminals) of SCR "A."

7. Your meter should indicate a very high or infinite resistance. A low resistance reading indicates a faulty SCR. Replace the module.

8. While leaving your ohmmeter connected as in steps 5 and 6 above, use a clip lead to connect the gate of SCR "A" (small terminal at the front of the module) to the red (positive) lead of your ohmmeter.

9. You should notice a drop in the resistance reading. This is a result of your triggering SCR "A" into conduction. If your resistance reading does not drop, the SCR may be faulty and you should replace the module. However, it is possible that your ohmmeter is not supplying sufficient current to trigger the SCR into conduction, and the SCR may be functioning normally. The old saying, "If in doubt, change it out!" would apply.

If SCR "A" appears normal, repeat the same process to check SCR "B:"

1. Place the black (negative) lead of your ohmmeter on the anode (terminal closest to the gate terminals) of SCR "B."

2. Place the red (positive) lead of your ohmmeter on the cathode (terminal farthest from the gate terminals) of SCR "B."

3. Your meter should indicate a very high or infinite resistance. A low resistance reading indicates a faulty SCR. Replace the module.

4. Reverse your meter connections. Place the black (negative) lead of your ohmmeter on the cathode (terminal farthest from the gate terminals) of SCR "B."

5. Place the red (positive) lead of your ohmmeter on the anode (terminal closest to the gate terminals) of SCR "B."

6. Your meter should indicate a very high or infinite resistance. A low resistance reading indicates a faulty SCR. Replace the module.

7. While leaving your ohmmeter connected as in steps 5 and 6 above, use a clip lead to connect the gate of SCR "B" (small terminal at the back of the module) to the red (positive) lead of your ohmmeter.

8. You should notice a drop in the resistance reading. This is a result of your triggering SCR "B" into conduction. If your resistance reading does not drop, the SCR may be faulty and you should replace the module. However, it is possible that your ohmmeter is not supplying sufficient current to trigger the SCR into conduction, and the SCR may be functioning normally. Again, "If in doubt, change it out!"

Figure 14-3 shows a transistor power module. Notice that this module has seventeen terminals, five large and twelve small. Figure 14-4 shows the schematic diagram of the transistor power module. This module may look intimidating at first glance, but notice that the module consists simply of six transistors connected to perform an inverter function.

Looking at both Figure 14-3 and Figure 14-4, notice the row of six small terminals at the top of the module. Three of these terminals (BU, BV, and BW) are connected to the bases of transistors Q1, Q2, and Q3. The remaining three terminals (EU, EV, and EW) are connected to the emitters of the same transistors. Now notice the row of six small terminals at the bottom of the module. Three of these terminals (BX, BY, and BZ) are connected to the bases of transistors Q4, Q5, and Q6. The remaining three terminals (EX, EY, and EZ) are connected to the emitters of the same transistors. You will also see two large terminals at the left end of the module. The large terminal toward the top of the module is the positive (+) DC input to the module. This terminal is internally connected to the collectors of transistors Q1, Q2, and Q3. The large terminal toward the bottom of the module is the negative (–) DC input to the module. This terminal is internally connected to the emitters of transistors Q4, Q5, and Q6. Finally, you should see three large terminals in the center of the module (between the top

Figure 14-3: *Transistor power module.*

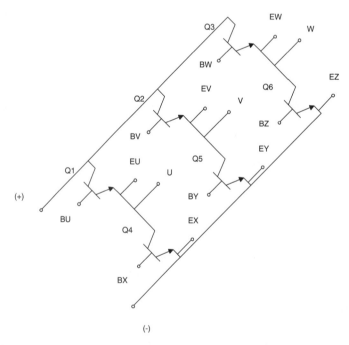

Figure 14-4: *Transistor power module schematic.*

and bottom rows of small terminals). The large terminal (W) at the right side of the module is connected to the emitter of Q3 and the collector of Q6. The next large terminal (V) to the left is connected to the emitter of Q2 and the collector of Q5. The last large terminal (U) to the left is connected to the emitter of Q1 and the collector of Q4.

You can perform a reasonably accurate test of this module with an ohmmeter. You should be aware, however, that often the circuitry inside the module is more complex than what is shown in Figure 14-3. The additional, but unknown components may cause erroneous or unexpected readings on the ohmmeter. Therefore, you will need to exercise some judgment in interpreting the ohmmeter readings to determine if the module is faulty or not.

To test the module with an ohmmeter, proceed as follows:

1. Place the ohmmeter in the "Ohms" position, if using a digital multimeters. (Use the 200 Ω position if using an analog multimeters.)

2. Place the black (negative) lead of your ohmmeter on the base (terminal BU) of transistor "Q1."

3. Place the red (positive) lead of your ohmmeter on the emitter (terminal EU) of transistor "Q1."

4. Your meter should indicate a very high or infinite resistance. A low resistance reading indicates a faulty transistor. Replace the module.

5. Leave the black (negative) lead of your ohmmeter connected to the base (terminal BU) of transistor "Q1."

6. Place the red (positive) lead of your ohmmeter on the (+) terminal of the module. (This connects your red lead to the collector of transistor "Q1.")

7. Your meter should indicate a very high or infinite resistance. A low resistance reading indicates a faulty transistor. Replace the module.

8. Reverse your meter connections. Place the black (negative) lead of your ohmmeter on the emitter (terminal EU) of transistor "Q1."

9. Place the red (positive) lead of your ohmmeter on the base (terminal BU) of transistor "Q1."

10. Your meter should indicate a low resistance. A high or infinite resistance reading indicates a faulty transistor. Replace the module.

11. Leave the red (positive) lead of your ohmmeter connected to the base (terminal BU) of transistor "Q1."

12. Place the black (negative) lead of your ohmmeter on the (+) terminal of the module. (This connects your black lead to the collector of transistor "Q1.")

13. Your meter should indicate a low resistance. A high or infinite resistance reading indicates a faulty transistor. Replace the module.

14. Place the red (positive) lead of your ohmmeter on the emitter (terminal EU) of transistor "Q1."

15. Place the black (negative) lead of your ohmmeter on the (+) terminal of the module. (This connects your black lead to the collector of transistor "Q1.")

16. Your meter should indicate a very high or infinite resistance. A low resistance reading indicates a faulty transistor. Replace the module.

17. Reverse your meter connections. Place the red (positive) lead of your ohmmeter on the (+) terminal of the module. (This connects your red lead to the collector of transistor "Q1.")

18. Place the black (negative) lead of your ohmmeter on the emitter (terminal EU) of transistor "Q1."

19. Your meter should indicate a very high or infinite resistance. A low resistance reading indicates a faulty transistor. Replace the module.

20. Repeat steps 1 through 19 for each of the remaining five transistors in the module. If any readings are questionable, remove any doubt by checking your measurements against a known good module, or simply replace the questionable module.

While we are on the subject of power modules, there is another type of power module that you may find in the drive upon which you are working. This module is technically not part of the power output section, although it does have a power function in the drive. The module is a three-phase bridge rectifier module. It is used to convert three-phase AC into rectified DC. A picture of this module appears in Figure 14-5 and a schematic of this module appears in Figure 14-6.

Notice that this module has five terminals. The two horizontal terminals at the left end of the module are the (+) and (–) DC connections. The three vertical terminals are the connections for the three-phase AC (L1, L2, and L3).

To test the module with an ohmmeter, follow these steps:

1. Place the ohmmeter in the "Diode Test" position, if using a digital multimeters. (Use the 200 Ω position if using an analog multimeters.)

2. Place the black (negative) lead of your ohmmeter on the (–) terminal of the module.

Figure 14-5: *Three-phase bridge rectifier module.*

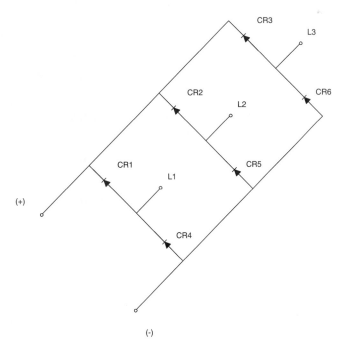

Figure 14-6: *Three-phase bridge rectifier module schematic.*

3. Place the red (positive) lead of your ohmmeter on AC terminal "L1" (this is actually the cathode of diode "CR4").

4. Your meter should indicate a very high or infinite resistance. A low resistance reading indicates a faulty diode. Replace the module.

5. Move the red (positive) lead of your ohmmeter from AC terminal "L1" to AC terminal "L2" (this is actually the cathode of diode "CR5").

6. Your meter should indicate a very high or infinite resistance. A low resistance reading indicates a faulty diode. Replace the module.

7. Move the red (positive) lead of your ohmmeter from AC terminal "L2" to AC terminal "L3" (this is actually the cathode of diode "CR6").

8. Your meter should indicate a very high or infinite resistance. A low resistance reading indicates a faulty diode. Replace the module.

9. Reverse your meter connections. Place the red (positive) lead of your ohmmeter on the (–) terminal of the module.

10. Place the black (negative) lead of your ohmmeter on AC terminal "L1" (this is actually the cathode of diode "CR4").

11. Your meter should indicate a low resistance. A high or infinite resistance reading indicates a faulty diode. Replace the module.

12. Move the black (negative) lead of your ohmmeter from AC terminal "L1" to AC terminal "L2" (this is actually the cathode of diode "CR5").

13. Your meter should indicate a low resistance. A high or infinite resistance reading indicates a faulty diode. Replace the module.

14. Move the black (negative) lead of your ohmmeter from AC terminal "L2" to AC terminal "L3" (this is actually the cathode of diode "CR6").

15. Your meter should indicate a low resistance. A high or infinite resistance reading indicates a faulty diode. Replace the module.

16. Place the red (positive) lead of your ohmmeter on the (+) terminal of the module.

17. Place the black (negative) lead of your ohmmeter on AC terminal "L1" (this is actually the anode of diode "CR1").

18. Your meter should indicate a very high or infinite resistance. A low resistance reading indicates a faulty diode. Replace the module.

19. Move the black (negative) lead of your ohmmeter from AC terminal "L1" to AC terminal "L2" (this is actually the anode of diode "CR2").

20. Your meter should indicate a very high or infinite resistance. A low resistance reading indicates a faulty diode. Replace the module.

21. Move the black (negative) lead of your ohmmeter from AC terminal "L2" to AC terminal "L3" (this is actually the anode of diode "CR3").

22. Your meter should indicate a very high or infinite resistance. A low resistance reading indicates a faulty diode. Replace the module.

23. Reverse your meter connections. Place the black (negative) lead of your ohmmeter on the (+) terminal of the module.

24. Place the red (positive) lead of your ohmmeter on AC terminal "L1" (this is actually the anode of diode "CR1").

25. Your meter should indicate a low resistance. A high or infinite resistance reading indicates a faulty diode. Replace the module.

26. Move the black (negative) lead of your ohmmeter from AC terminal "L1" to AC terminal "L2" (this is actually the anode of diode "CR2").

27. Your meter should indicate a low resistance. A high or infinite resistance reading indicates a faulty diode. Replace the module.

28. Move the black (negative) lead of your ohmmeter from AC terminal "L2" to AC terminal "L3" (this is actually the anode of diode "CR3").

29. Your meter should indicate a low resistance. A high or infinite resistance reading indicates a faulty diode. Replace the module.

Once you have completed testing the module, and you determine that the module is defective, you can usually obtain a substitute part from a local electronics parts supplier. If the part is not available, you will have to return the drive to the manufacturer for service or call a service technician for on-site repairs.

If it is determined that the problem is not in the power section, then it must be located in the control section of the drive. The electronics used in the control section are more complex, and therefore troubleshooting is not recommended. In this event, the drive must be returned to the manufacturer for repair or arrangements made for on-site repair by a factory trained technician.

Review Questions

1. Name the four main areas to check when a problem occurs with an AC inverter drive.

2. Where is the best place to begin troubleshooting an AC inverter drive? Why?

3. List some safety steps you should follow prior to working on an AC inverter drive.

4. Is it permissible to connect an AC inverter to a supply voltage that is higher than the nameplate rating of the drive? Why or why not?

5. Explain phase imbalance.

6. Calculate the phase imbalance for a supply to an AC inverter that has the following voltages: L1 to L2 = 231 V; L2 to L3 = 241 V; and L1 to L3 = 243 V.

7. **True or False?** An inoperable cooling fan can prevent an inverter from operating. Explain your answer.

8. Describe the cooling problem caused by using an inverter drive to operate a motor with a shaft-mounted fan.

9. Describe the dangers of using a megger on an AC inverter drive system.

10. Explain how the motor load can affect the performance of the AC inverter drive.

11. **True or False?** When using shielded cable on feedback devices, you must ground the shield wire at both ends of the cable. Justify your answer.

12. Describe any precautions that should be observed when routing the feedback device cable outside the drive cabinet.

13. Describe any precautions that should be observed when routing the feedback device cable inside the drive cabinet.

14. Explain when it is permissible to readjust the settings on the drive.

15. Explain when an AC inverter should be returned to the manufacturer for repair.

Appendix A

DC Motor Review

The typical DC motor consists of four main parts; the armature, the commutator, the brush assembly, and the field windings. The armature typically has loops of wire mounted on a rotating shaft. The ends of these loops are connected to the commutator. The commutator provides a surface on which the brush assembly makes electrical contact, allowing current to be applied through the assembly to the revolving armature windings. The field windings, which are also connected to a source of electrical power, do not revolve but instead remain stationary. Figure A-1 shows the previously mentioned parts of a DC motor.

The purpose of the commutator is to reverse the polarity of the DC voltage applied to the rotor. This causes the rotor to turn only in one direction because

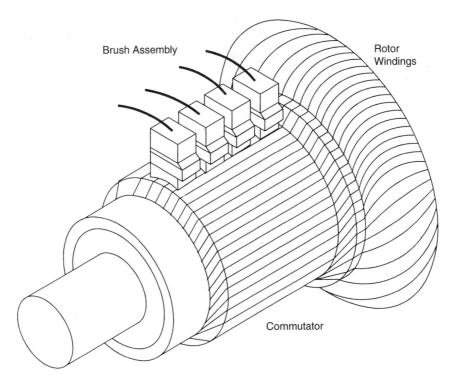

Brush Assembly

Rotor
Windings

Commutator

Figure A-1: Parts of a DC motor.

of the attraction between the magnetic fields in the stator and rotor. Notice in Figure A-2 that the stator is connected to a DC source. Observe the polarity of the stator windings. Also, notice that the connections to the stator do not change. The stator will always have the same magnetic polarity. Now watch what happens as the rotor revolves. When DC is applied to the rotor in Figure A-2, the magnetic field of the rotor causes the rotor to be attracted to the stator winding with the opposite magnetic polarity. The rotor thus turns in the direction of stator windings. Now look at Figures A-3 through A-5. The polarity of the DC voltage that is applied to the rotor is reversed by the commutator, causing the rotor to continue to rotate in the same direction. The speed of the rotor is controlled by the strength of the magnetic field in the rotor, usually by means of a variable resistor in the rotor circuit. To change the direction of the rotor's rotation, we can reverse the DC connections either to the rotor or to the stator.

DC motors can be connected in one of four ways. A motor connected as shown in Figure A-6 is called a series wound motor because the **series field** is connected in series with the armature. The series wound motor produces high torque at low speeds. Unfortunately, it does not possess good speed regulation. In fact, a loss of load creates a serious problem in the series wound DC motor. If the load is removed, the series wound DC motor speeds up. The motor's speed can increase to such a high value that the motor will actually fly apart because of the centrifugal forces that high speeds generate. For this reason, series

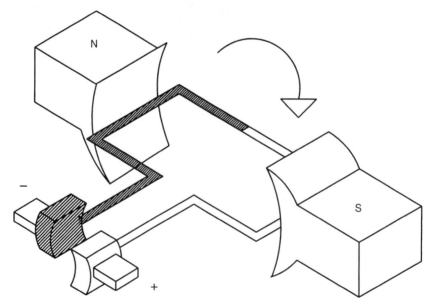

Figure A-2: *Commutator at 0°.*

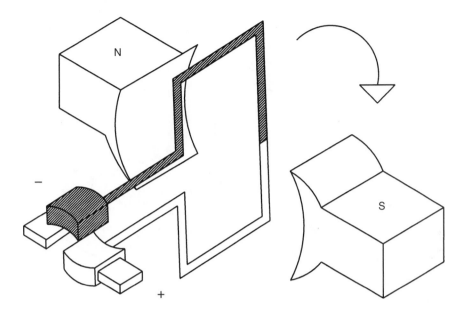

Figure A-3: Commutator at 90°.

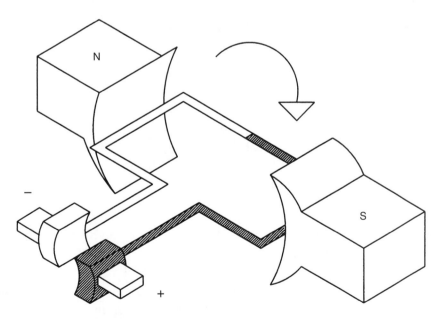

Figure A-4: Commutator at 180°.

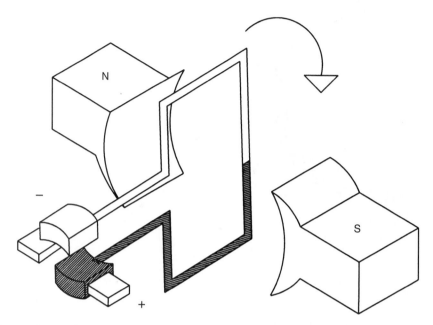

Figure A-5: *Commutator at 270°.*

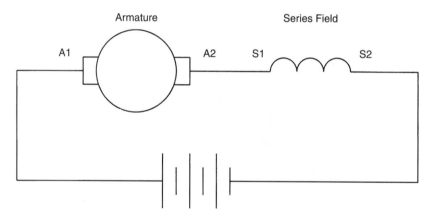

Figure A-6: *Series wound DC motor.*

wound DC motors should never be used with belt- or chain-driven loads. These motors should always be directly coupled to their loads by shafts.

A second way to connect a DC motor is to connect the field windings across or in parallel with the armature. This configuration, known as a shunt wound DC motor, is shown in Figure A-7. In this circuit the shunt winding receives power from the same source as the armature.

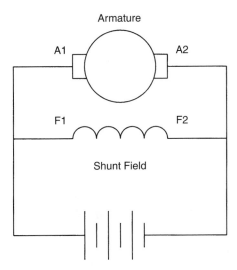

Figure A-7: *Shunt wound DC motor.*

By changing the preceding circuit slightly, we arrive at the third method of connecting a DC motor, illustrated in Figure A-8. Notice that the shunt winding now receives power from a separate DC source. This motor is called a separately excited shunt wound DC motor. (If the DC motor receives its **excitation current** from the motor's self-generated DC, the motor is known as a **self-excited DC motor.**) Most DC motors are connected in this manner when used with DC drives. The DC shunt motor offers better speed regulation than the series wound motor. Also, the shunt motor will not speed up and fly apart if the load is removed. However, if the shunt field winding develops an open circuit, the shunt motor will speed up and can fly apart. For this reason, most installations include a field loss relay that removes power from the motor when it detects an open field condition.

The final method of connecting a DC motor is the compound wound DC motor configuration shown in Figure A-9 (see next page). This motor has a combination of features of the series and the shunt wound DC motors. Notice that a series field is connected in series with the armature and a shunt field is connected in parallel with the armature. This arrangement is the reason why the motor is called a compound wound motor. The compound wound motor combines the performance characteristics of both motor types as well. Its speed regulation is not as good as that of the shunt wound motor, but better than that of the series wound motor. The starting torque is better than that of the shunt wound motor, but not as good as that of the series wound motor.

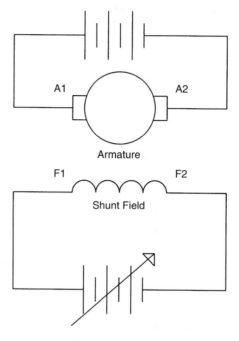

Figure A-8: *Separately excited shunt wound DC motor.*

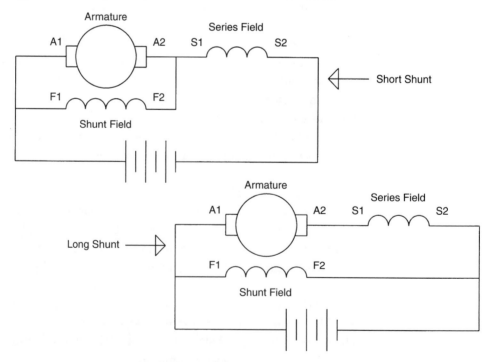

Figure A-9: *Compound wound DC motor.*

160

Appendix B

AC Motor Review

The various types of AC motors can be divided into two main categories: single-phase motors and three-phase motors. These two categories can be further divided according to the various techniques by which they produce mechanical power from electrical energy. We will discuss only one type of single-phase motor and one type of three-phase motor in this section.

We begin with one of the most widely used single-phase motors, the capacitor start/capacitor run motor shown in Figure B-1. Notice the various parts of the capacitor start/capacitor run motor. This motor has a rotor, two capacitors, and a stator that consists of a start winding and a run winding.

Notice that the rotor, illustrated in Figure B-2 on page 162, has windings. Instead, the rotor consists of metal bars connected at each end to end rings. Between the metal bars are sheets of laminated metal. During the motor's operation, voltage is induced into the metal bars, producing current flow and a magnetic field.

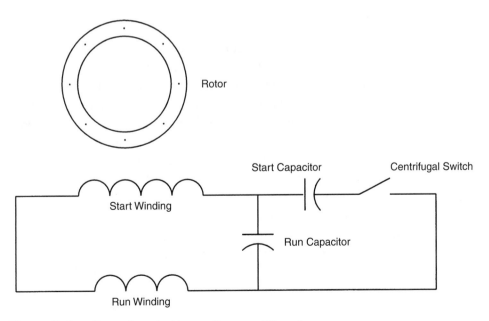

Figure B-1: Capacitor start/capacitor run AC motor.

Figure B-2: *Squirrel cage type rotor.*

Now we will consider the stator in Figure B-1. The stator consists of two separate windings: the run winding and the start winding. The start winding has many turns of large-gauge wire, whereas the run winding has fewer turns of smaller gauge wire. Notice also that a capacitor is connected in series with the start winding by means of a centrifugal switch. The combination of the capacitor and the start winding causes the starting current to undergo a phase shift that typically approaches 90°. Observe that the start winding and the run winding are physically located 90 mechanical degrees apart in the stator housing. These windings are always connected to the supply and together produce the effect of a rotating magnetic field in the stator. The rotor is attracted to this field and begins to revolve. As a result of the high phase shift, the capacitor start/capacitor run motor has a high starting torque.

To allow the motor to maintain a high running torque, a centrifugal switch permits the run capacitor to remain in the circuit after the motor has attained a preset speed. The following series of events accomplishes this. As the motor's speed increases, centrifugal forces cause the switch to open, removing the start

Figure B-3: *Frame, stator windings, and rotor of a squirrel cage motor. End bell removed.*

capacitor from the circuit. The switch has no effect on the run capacitor, however. As the motor slows to a stop, the switch closes once the centrifugal forces diminish to a certain level. This event automatically switches the start capacitor back into the circuit for restarting.

In the past, the speed of the capacitor start/capacitor run motor was considered fixed. Today, thanks to inverters, we are able to vary the frequency of the power applied to the motor, and thus its speed as well.

Now we turn our attention to what is probably the most widely used motor in industry today: the three-phase squirrel cage induction motor. Notice that the squirrel cage motor in Figure B-3 has two parts: the rotor and the three-phase stator. In fact, this simple design is one of the squirrel cage motor's most attractive features. Let's see how this motor works.

Notice that the squirrel cage motor has the same type of rotor as the capacitor start/capacitor run motor. In fact, it is for this rotor, which resembles the

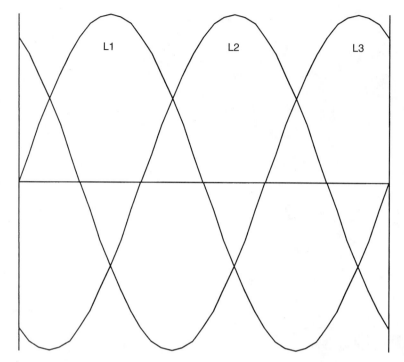

Figure B-4: *Three-phase AC.*

exercise wheel commonly found in mouse and hamster cages, that the squirrel cage motor is named.

Figure B-4 illustrates the three-phase AC power applied to the stator windings. Recall that three-phase power produces a peak voltage every 120°. Now notice the three separate phase windings in the stator in Figure B-5. Notice also that the phase windings are arranged in sequence around the stator housing. As the applied three-phase power peaks in the positive direction in phase L1, phase windings L2 and L3 will have opposite polarity, and their value will be between 0 and their maximum negative value. As phase L2 peaks in its positive direction, L1 and L3 will have opposite polarity, and their value will be between 0 and their maximum negative value. This pattern continues through the entire 360° rotation of the applied AC sine wave.

If you concentrate on the stator windings, you will see that a rotating magnetic field appears around the stator. This magnetic field induces a voltage into the rotor bars, causing a current flow. The current flow in the rotor bars pro-

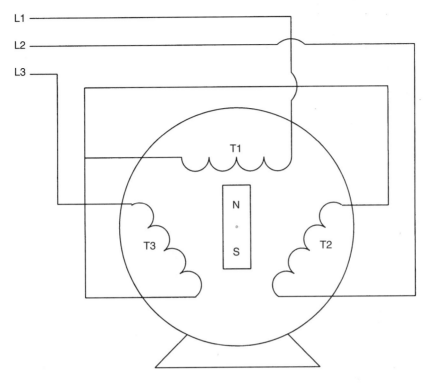

Figure B-5: *Three-phase stator windings.*

duces another magnetic field that is attracted to the revolving magnetic field in the stator. This attraction causes the rotor to turn and produce a torque.

Like the capacitor start/capacitor run motor, the squirrel cage motor was once considered a fixed speed motor. Today, however, with inverters we can vary the frequency of the applied AC power, and consequently, the speed of the squirrel cage motor as well.

Appendix C

Power Semiconductors

Many types of power semiconductors are on the market today, and the list grows continuously. Some of the most common types used in the electronic variable speed drive industry today are the P-N junction diode, thyristors, bipolar transistors, MOSFETs, and insulated gate bipolar transistors.

The P-N Junction Diode

The **P-N junction diode** in Figure C-1 is a two-terminal device. One terminal, the anode, is represented by the arrowhead symbol. The other terminal, the cathode, is represented by the T-shaped terminal.

You can perform a simple test on a diode, either in or out of the circuit, to determine if the diode is functioning properly. When testing in circuit, be certain that power is removed and that any capacitors are fully discharged. If any of the readings from an in-circuit test are questionable, remove the diode from the circuit and retest it out of circuit.

Performing the test requires nothing more than a simple ohmmeter. To see how the test works, refer to Figure C-2 on the next page. Most digital multimeters (DMM) have a diode check position, indicated by the diode symbol. Set the meter to this position. Then, place the negative lead of the DMM on one terminal of the diode to be tested and connect the positive lead from the DMM to the other terminal of the diode. Notice the reading on the DMM. If the diode is connected properly (negative to the cathode and positive to the anode) and is working properly, the DMM will display a reading of a few tenths of a volt.

There are several reasons why the DMM may not display this reading. The diode may be defective; you may have connected the diode backwards (reverse biased); or the DMM may be defective. You can verify whether the DMM is working properly by checking a known good diode. Switch the connections

Cathode   Anode

Figure C-1: Schematic symbol representing the P-N junction diode.

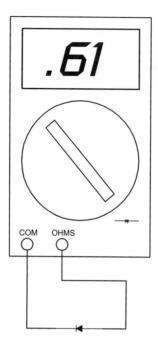

Figure C-2: *Forward biasing a diode with an ohmmeter.*

between the DMM and the diode you are testing. This will check the diode for conduction in the opposite direction from your original settings. If the DMM now reads several tenths of a volt, the diode is forward biased, as it should be, and is probably good. In the process, you have also identified its cathode (the terminal connected to the negative lead of the DMM) and its anode (the terminal connected to the positive lead of the DMM). If the DMM does not produce a reading with the diode connected either way, the diode is probably open and should be replaced. If the DMM produced the same reading with the diode connected either way, the diode is probably shorted and should be replaced. Even if the diode proved to be in good condition, it still may fail under circuit conditions. Although this is a remote possibility, it is a possibility nonetheless.

Diodes are available in many different case styles, or packages, some of which are shown in Figure C-3.

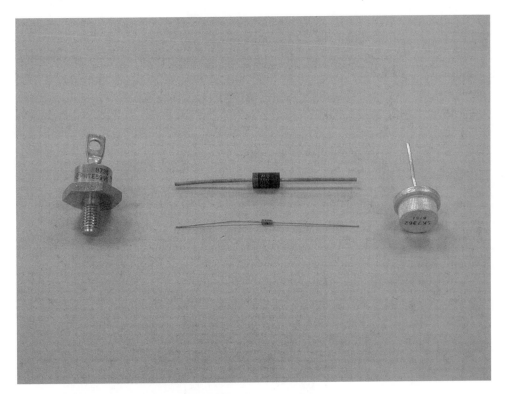

Figure C-3: *Various diode packages.*

Thyristors

Thyristors, more commonly called **silicon controlled rectifiers (SCRs),** are three-terminal devices. Notice in Figure C-4 (see page 170) that the SCR looks very similar to a diode except that it has an extra terminal. The terminals of the SCR are the anode (represented by the arrowhead), the cathode (represented by the T-shaped symbol), and the gate or trigger (the remaining terminal in the illustration).

To see how to perform a simple test with a DMM to determine whether an SCR is functioning, refer to Figure C-5 on page 170. This test may be performed either in or out of circuit. However, when testing in-circuit, be certain that the power is disconnected and that any capacitors are fully discharged. If any of the readings are questionable, the SCR should be removed from the circuit and retested out of circuit.

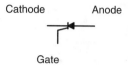

Cathode   Anode

Gate

Figure C-4:   *Schematic symbol representing the silicon controlled rectifier (SCR).*

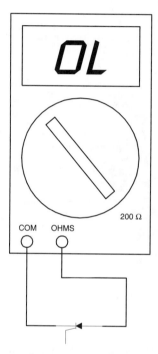

Figure C-5:   *Forward biasing an SCR with an ohmmeter.*

Begin by testing the diode portion of the SCR. Set the DMM either to the 200 Ω or to the 2 kΩ positions. Do not use the low-Ω or diode check positions, as these will not supply sufficient current to check the SCR under test accurately. (If you are using an analog multimeter, set the meter to the "Ohms" position. Use the R × 10, R × 100, or R × 1 kΩ ranges. Other ohmmeter ranges may produce damaging voltage levels and erroneous readings.) Connect the negative lead from the DMM to the cathode of the SCR you are testing. Connect the positive lead from the DMM to the anode of the SCR. A good SCR will typically measure in excess of 100 kΩ. Now reverse your connections. Connect the negative lead from the DMM to the anode of the SCR and the positive lead from the DMM to the

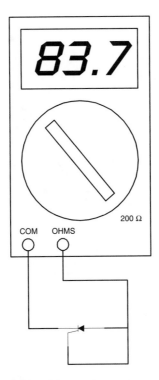

Figure C-6: *Triggering an SCR with an ohmmeter.*

cathode of the SCR. You should still obtain a reading in excess of 100 kΩ if the SCR is good.

Now refer to Figure C-6 and reverse your connections again, so that the negative lead from the DMM is connected to the cathode of the SCR and the positive lead from the DMM is connected to the anode of the SCR. Next take another test lead and, leaving the DMM connected as described, connect this third lead from the positive lead of the DMM (at the anode) to the gate terminal of the SCR. (Thus the positive lead of the DMM is now connected both to the anode and to the gate of the SCR.) If the SCR is good, the resistance reading should drop to less than 1 kΩ. If there is no change from the previous resistance reading, the SCR is probably defective and will need to be replaced.

If you remove the connection to the gate terminal, the resistance reading should remain at less than 1 kΩ. However, if the reading returns to a higher resistance value, the SCR is not necessarily defective. Some DMMs do not supply sufficient holding current for the SCR to maintain conduction. This is especially true with larger, high-current SCRs.

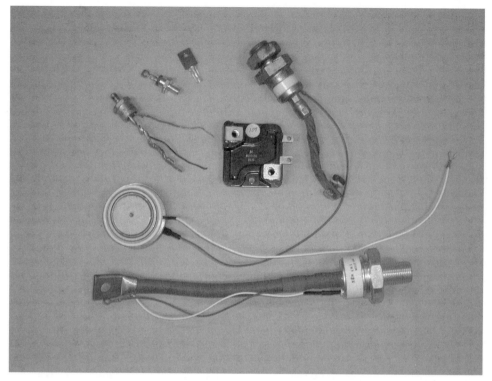

Figure C-7:   *Various SCR packages.*

SCRs are available in many different case styles, or packages, some of which are shown in Figure C-7.

The Bipolar Junction Transistor

The **bipolar junction transistor (BJT),** a three-terminal device, is available in the two versions shown in Figure C-8 on page 173. Notice that the directions of the arrowhead on the two terminals are different. The symbol on the right with the arrowhead pointing inward is known as a **PNP transistor.** The symbol with the arrowhead pointing outward is known as an **NPN transistor.** The theory of operation of both types of transistors is identical; only the direction of current flow through the two devices differs.

The three terminals of a bipolar junction transistor are: the emitter (represented by the arrowhead), the base (the T-shaped terminal), and the collector. Some bipolar junction transistors have only two leads, as shown in Figure C-9.

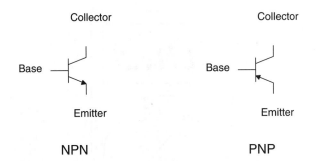

Figure C-8: *Schematic symbols representing the NPN and PNP bipolar junction transistor.*

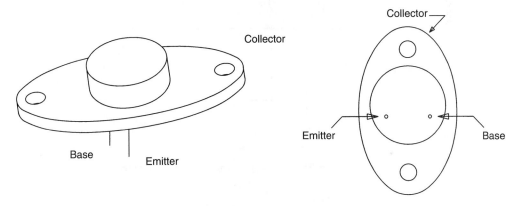

Figure C-9: *Metal body transistor with two terminals.*

In this instance the collector is usually the metal body of the transistor itself. Next, we will see how to test a bipolar junction transistor with a DMM.

The transistor may be tested in or out of circuit. When testing in circuit, remember to de-energize the circuit and verify that any capacitors are fully discharged. Note that if a transistor proves questionable when tested in circuit, you should retest it out of circuit to be certain whether it is defective.

We begin by setting the DMM to a low resistance range, such as 200 Ω or 2 kΩ. Do not use the low-Ω or diode check positions, as these do not supply sufficient current to test the transistor accurately. (If you are using an analog multimeter, set the meter to the "Ohms" position. Use the R × 10, R × 100, or R × 1 kΩ ranges. Other ohmmeter ranges may produce damaging voltage levels and erroneous readings.)

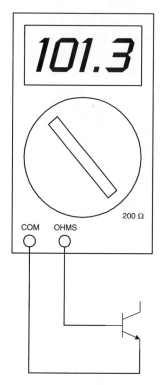

Figure C-10:   *Testing the emitter-base junction with an ohmmeter.*

Notice how the NPN transistor in Figure C-10 is connected to the DMM. The negative lead of the DMM is connected to the emitter of the transistor. The positive lead of the DMM is connected to the base of the transistor. If the transistor is good, the DMM reading will reflect a low resistance (typically less than 1 kΩ).

Next, reverse the DMM connections. Put the positive lead of the DMM on the emitter and the negative lead from the DMM on the base. The resulting reading from a good transistor will typically measure in excess of 100 kΩ. If you measured a low value in both instances, the emitter-base junction is probably shorted. If the measurement was high in both instances, the emitter-base junction is probably open. In either of these two cases the transistor should be replaced.

If the emitter-base junction test proves this area is functioning properly, you should check the collector-base junction next, shown in Figure C-11. First, connect the negative lead of the DMM to the collector of the transistor. Then connect the positive lead of the DMM to the base of the transistor. If the collector-base

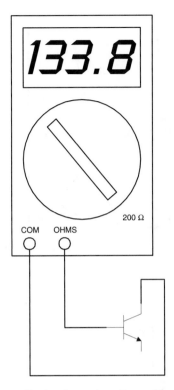

Figure C-11: Testing the collector-base junction with an ohmmeter.

junction is good, the reading should measure less than 1 kΩ. Now reverse the connections to the transistor. Connect the positive lead from the DMM to the collector of the transistor and the negative lead of the DMM to the base of the transistor. If the collector-base junction is good, the reading should measure greater than 100 kΩ. If the reading was a low value in both instances, the collector-base junction is probably shorted. If the measurement was high in both instances, the collector-base junction is probably open. In either case the transistor should be replaced.

    If the collector-base junction test results prove that this area of the transistor is functioning properly, you may proceed to perform the final check, involving the emitter-collector junction as shown in Figure C-12. Begin by connecting the negative lead of the DMM to the emitter terminal. Next connect the positive lead of the DMM to the collector. If the transistor is good, the DMM reading will be greater than 100 kΩ.

    Next, reverse the connections to the transistor. Connect the positive lead of the DMM to the emitter and the negative lead of the DMM to the collector.

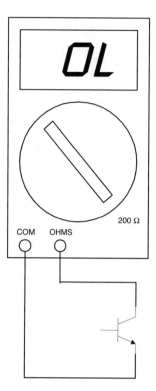

Figure C-12: *Testing the emitter-collector junction with an ohmmeter.*

Again, the DMM reading should be greater than 100 kΩ. If the DMM indicated a low resistance in both tests, the transistor is probably shorted and should be replaced.

If the transistor passed all of the above tests, it is more than likely functioning properly. However, under normal operating conditions of circuit voltage, the transistor may still possibly break down and fail. It is also common for transistors to fail after they have heated up. Therefore the DMM test is not absolutely reliable. Nevertheless it can be an effective and quick troubleshooting tool.

Bipolar junction transistors are available in many different case styles, or packages, several of which are shown in Figure C-13.

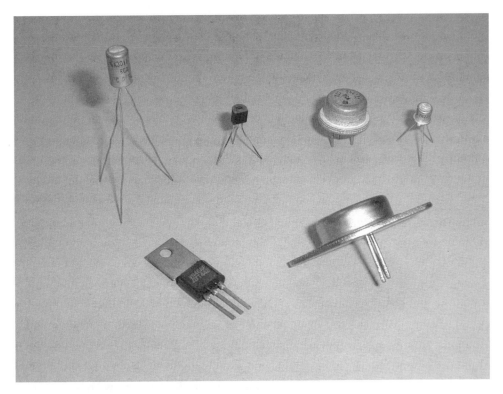

Figure C-13:   *Various transistor packages.*

MOSFETs

A **MOSFET** is a three-terminal device, as shown in Figure C-14. The terminals are: the source (represented by the arrowhead), the Gate (represented by the L-shaped terminal), and the drain.

Due to its nature, a MOSFET does not lend itself to reliable testing with an ohmmeter. In-depth understanding of electronic circuits is needed to determine if a MOSFET is defective. Also, because most MOSFETs are sensitive to static electricity, special handling precautions are required to prevent static discharge from damaging these devices. Troubleshooting is therefore not recommended. If you suspect that a MOSFET is the cause of a drive's failure, you should refer servicing of the drive to a qualified technician or return the drive to the manufacturer for repair.

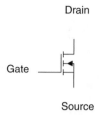

Figure C-14: *Schematic symbol representing the metal oxide semiconductor field effect transistor (MOSFET).*

Insulated Gate Bipolar Transistors

The **insulated gate bipolar transistor (IGBT)** is another three-terminal device. Two different schematic symbols are used to represent the IGBT (see Figures C-15 and C-16). In Figure C-15 the terminal represented by the arrowhead pointing outward is the emitter, and the terminal with the arrowhead pointing inward is the collector. The L-shaped terminal is the gate. In Figure C-16 the emitter is the terminal represented by the arrowhead. Again, the L-shaped lead is the gate. The remaining lead is the collector.

Because of its structure, the IGBT cannot be reliably tested with an ohmmeter. In-depth understanding of electronic circuits is needed to determine if an IGBT is defective. Therefore, if you suspect that an IGBT is the cause of a drive's failure, you should refer servicing to a qualified field service technician, or return the drive to the manufacturer.

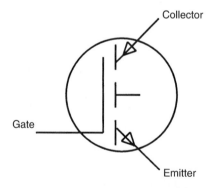

Figure C-15: *Schematic symbol representing the insulated gate bipolar transistor (IGBT).*

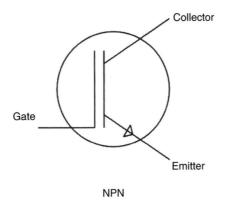

NPN

Figure C-16:   *Alternate schematic symbol representing the insulated gate bipolar transistor (IGBT).*

Appendix D

Manufacturers' Schematics

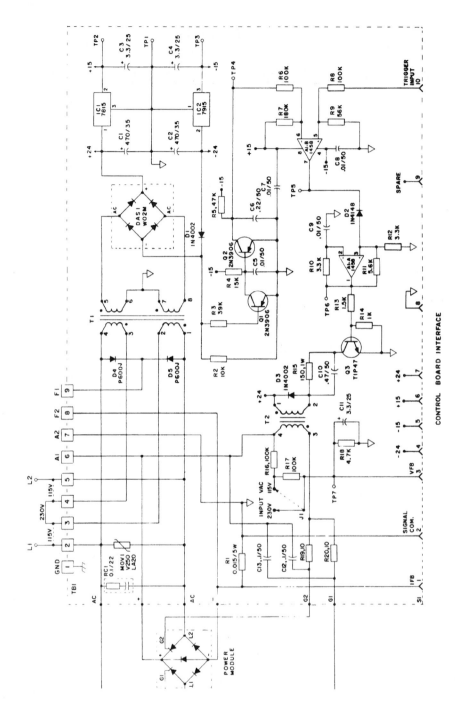

Figure D-1: Schematic of the power board for an ADP100 series DC drive. (Courtesy of Carotron, Inc.)

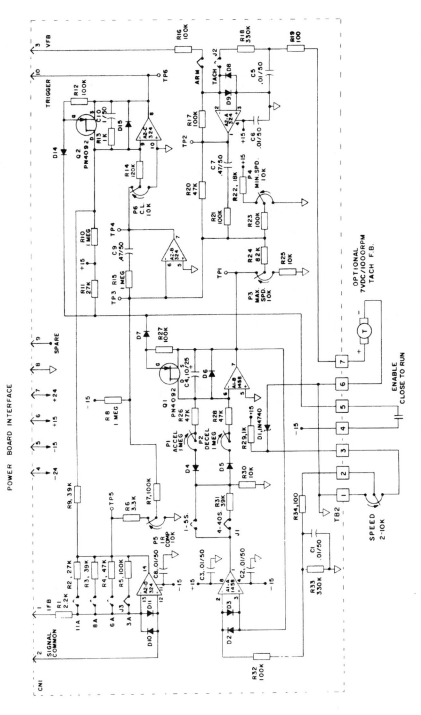

Figure D-2: Schematic of the speed control board for an ADP100 series DC drive. (Courtesy of Carotron, Inc.)

183

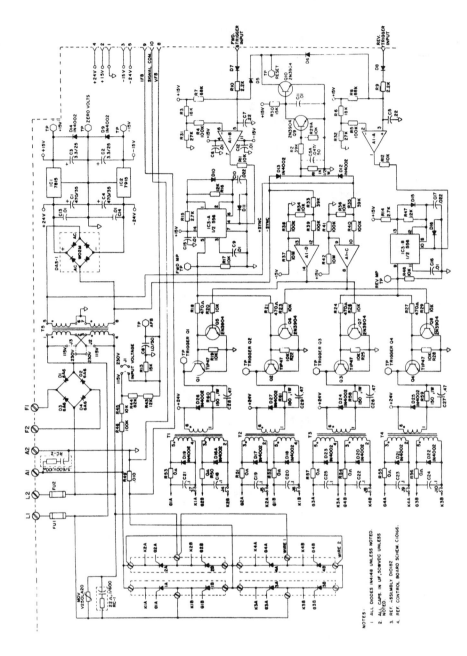

Figure D-3: Schematic of the power board for an RCP200 series DC drive. (Courtesy of Carotron, Inc.)

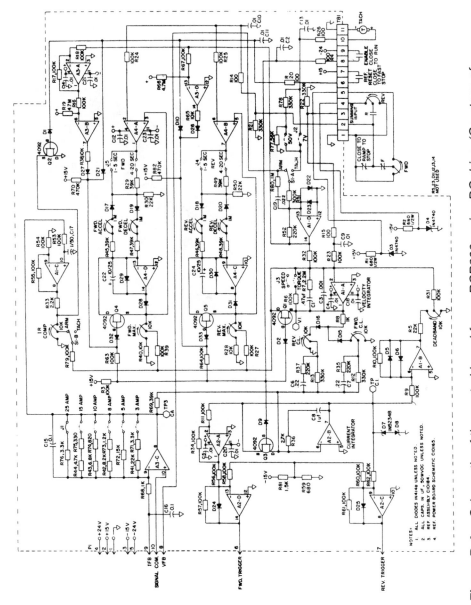

Figure D-4: Schematic of the control board for an RCP200 series DC drive. (Courtesy of Carotron, Inc.)

Appendix E

The Saturable Reactor

The main purpose of this text is to introduce the theory behind electronic variable speed drives. Although another type of variable speed drive exists which may arguably be classified as "electronic," discussion of it has been deferred until now because some do not consider it to be truly "electronic" in nature. This device, the saturable reactor, was widely used prior to the advent of the electronic variable speed drives discussed earlier in the text. What a saturable reactor is and how it works is the subject of this appendix.

As you studied DC and AC motors, you learned that motor speed can be varied by varying the voltage applied to the motor terminals. However, this very simple approach to motor speed control is not sufficiently precise to meet the demands of industry today. Nevertheless, if precise speed control is not needed, a saturable reactor can provide effective and inexpensive motor speed control. First we will discuss how a saturable reactor is constructed.

Notice in Figure E-1, that a *saturable reactor* appears very similar in construction to a typical transformer. There are some differences between them, however. First notice the reactor windings, or load windings, on the outside of the reactor core. These windings are specially constructioned to be equal in wire size and number of turns. For a given voltage and frequency, the two reactor windings therefore have equal inductive reactance and equal voltage drops. Actually, these windings are constructed to produce a high inductive reactance, which in turn causes a large voltage drop to develop across each reactor winding.

Also notice in Figure E-1 on the next page another winding located around the center of the reactor core. This *control winding* is separately excited when connected to a variable DC supply. Applying DC to this winding produces a magnetic flux that increases as the applied DC increases. However, a point is eventually reached at which increasing the applied DC does not produce a further increase in magnetic flux. When this point is reached, we say that the core is *saturated*. This phenomenon allows us to control or vary motor speed. Next, we will discuss in detail how the saturable reactor works.

We begin by examining how a saturable reactor may be used to drive an AC motor. Notice in Figure E-2 at the bottom of the next page that each reactor

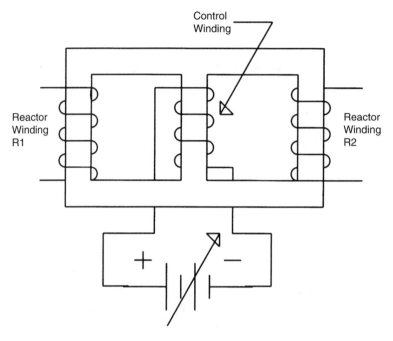

Figure E-1:   *Schematic of a saturable reactor.*

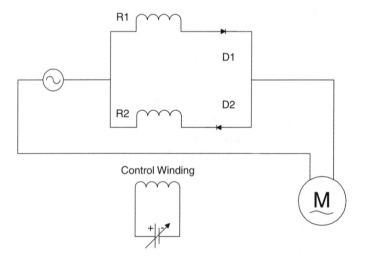

Figure E-2:   *Using a saturable reactor with an AC motor.*

winding is connected in series to a steering diode, and the two reactor winding/steering diode series combinations are connected in parallel to one another. However, observe that the polarities of the diodes are opposite. If we apply a source of AC to the AC input terminals, during one half-cycle current flows through the series combination consisting of reactor winding R1, steering diode D1, and the motor load. During the next half-cycle of the applied AC, the current flows through the series combination of reactor winding R2, steering diode D2, and the motor load. The motor does not operate, however. Remember that the reactor windings have a very large inductive reactance and thus function as a voltage divider. The resulting large drop in voltage across the reactor windings leaves very little, if any, voltage for the load.

To make the motor operate, we apply a DC control voltage to the control winding of the saturable reactor. Recall that this DC control voltage is variable.

Increasing the DC control voltage to the control winding creates a magnetic flux in the core of the saturable reactor. The amount of flux produced is proportional to the amount of DC control voltage applied to the control winding. Increasing the amount of flux causes the reactance of the reactor coils to decrease, and as a result less voltage is dropped across the coils. As the voltage drop across the coils decreases, the voltage available to the motor increases, causing the motor to operate at a low speed. Further increasing the amount of DC control voltage applied to the control winding makes more voltage available at the motor terminals and causes the motor speed to increase.

This process can be continued until the point is reached where the flux produced by the DC control voltage is so great that the iron core of the saturable reactor becomes saturated. At that point any further increase in DC control voltage will have no effect on the motor's speed. However, by lowering and raising the amount of DC control voltage we can cause the motor's speed to decrease or increase accordingly. Most saturable reactors are manufactured so that the ratio of control winding turns to reactor winding turns is large to allow control of very large amounts of AC output power with very small amounts of DC control power. For this reason, saturable reactors are also known as *magnetic amplifiers*.

Now we will see how to use a saturable reactor with a DC motor. Notice in Figure E-3 on the next page that reactor windings R1 and R2 are again connected in series with steering diodes D1 and D2 respectively. Also notice the two additional diodes, D3 and D4. By now, you probably recognize the familiar full-wave bridge circuit!

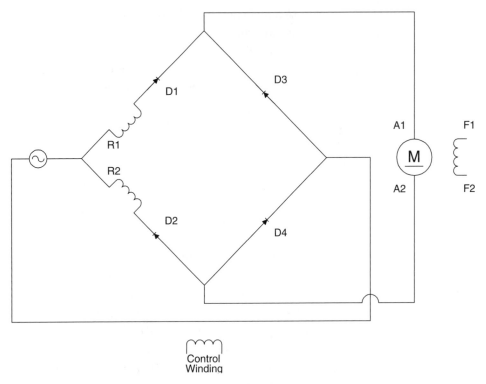

Figure E-3:   *Using a saturable reactor with a DC motor.*

Next we will consider how this configuration works. During the first half-cycle of the applied AC power, current flows through reactor winding R1, steering diode D1, the motor armature, and diode D4. During the next half-cycle, current flows through reactor winding R2, steering diode D2, the motor armature, and diode D3. Note that the direction of the current flow through the motor armature is the same for both halves of the applied AC power. This means that the motor armature is receiving DC power because the applied AC power has been rectified by diodes D3 and D4. This DC power is also variable and controlled. Just as in the case of the saturable reactor with AC output, we can use variable DC control voltage to control the DC output power in this circuit.

When no DC control voltage is applied to the saturable reactor, reactor windings R1 and R2 have a high reactance and drop a large amount of the applied voltage. As a result, very little, if any, DC voltage is applied to the armature. Now, applying a small amount of DC control voltage to the control winding of the saturable reactor produces a magnetic flux. This magnetic flux reduces both the reactance of the reactor windings and the voltage dropped across these windings. In turn, the voltage applied to the armature increases, as does the speed of the motor. Further increasing the DC control voltage applied to the control winding will cause the reactance of the reactor windings to decrease. This decreased reactance causes an increase in the voltage applied to the armature and a resulting increase in motor speed.

This process can continue until the saturable reactor reaches its saturation point. At which point any further increase in control voltage will have no effect on motor speed. As mentioned earlier, if the ratio of control winding turns to reactor winding turns is large, very large amounts of DC output power can thus be controlled with very small amounts of DC control power.

Glossary

analog input select Sets either a 0–10 V or 4–20 mA reference that is proportional to the frequency input.

analog output select Provides a 4–20 mA output signal that is proportional to the motor speed.

armature The portion of a motor or generator that usually rotates. The armature contains a laminated iron core with windings.

armature voltage feedback Another name for EMF feedback.

automatic control A control system in which a signal indicating motor performance is fed back to the control section of the drive so that the drive can take corrective action automatically. Also known as closed-loop control.

back EMF The induced voltage caused by the generator action of a revolving motor. Back EMF opposes the voltage applied to the motor and thereby limits the armature current. Also known as counter electromotive force (CEMF).

basic set-up As applied to drives, can be set for constant torque or constant speed.

BLDC motor Brushless DC motor.

bipolar junction transistor (BJT) A three-terminal, solid-state device. Its three terminals are the emitter, the base, and the collector. A bipolar junction transistor is a current-controlled device. The bipolar junction transistor can be used either as a switch or as an amplifier. Two types of bipolar junction transistors are available: NPN or PNP. Both types operate identically, except that the flow of current is reversed. Bipolar junction transistors are often referred to as BJTs or simply transistors.

BJT Bipolar junction transistor.

boost chopper A chopper circuit in which the output voltage is higher than the input voltage. Sometimes called a step-up chopper.

brush Typically, a block of carbon or graphite with a braided wire (called a pigtail) embedded inside. The brush provides an electrical path between the power supply and the commutator of a DC motor. As motor operating time increases, the brushes eventually wear out and need to be replaced.

brushless DC motor   A DC motor that uses electronics instead of a commutator and brush assembly to perform the commutation of the applied DC.

buck chopper   A chopper circuit in which the output voltage is lower than the input voltage. Sometimes called a step-down chopper.

CEMF   The induced voltage caused by the generator action of a revolving motor. CEMF opposes the voltage applied to the motor and thereby limits the armature current. Also known as counter electromotive force or back EMF.

chopper   A circuit used to break up or chop a steady DC into a series of pulses.

closed-loop control   A control system in which a signal indicating motor performance is fed back to the control section of the drive so that the drive can take corrective action automatically. Also known as automatic control.

command signal   The voltage level set or programmed into a drive to ensure that the desired motor operating speed is maintained. Also known as the reference signal or set point.

commutator   Copper segments or bars on the end of the shapft of a DC motor that provide an electrical path between the armature windings and the brushes. The commutator maintains proper current flow direction through the armature windings.

converter   A device that changes AC power into DC power.

counter electromotive force (CEMF)   The induced voltage caused by the generator action of a revolving motor. Counter electromotive force opposes the voltage applied to the motor and thereby limits the armature current. Also known as CEMF or back EMF.

CSI   Current source inverter. An inverter in which the DC bus current is controlled while the DC bus voltage varies to meet the demands of the motor. Also known as a current fed inverter.

current   As applied to drives, the output current of the inverter.

current fed inverter   An inverter in which the DC bus current is controlled while the DC bus voltage varies to meet the demands of the motor. Also known as a current source inverter or CSI.

current limit   Sets the maximum current available to the motor. If the setting is at the maximum permissible value, the motor will have maximum starting torque. The value can be on the order of 160% of nominal motor current. If the current limit is set too low, the inverter can trip out.

current source inverter An inverter in which the DC bus current is controlled while the DC bus voltage varies to meet the demands of the motor. Also known as a current fed inverter or CSI.

DC brake time If selected, this parameter will provide additional braking torque at low motor speeds.

DC filter Also called the DC link or DC bus. Provides smooth, rectified DC.

DC injection Also called plugging. Used to stop an AC motor by applying DC to the motor windings. This DC current replaces the rotating magnetic field with a fixed magnetic field. The rotor becomes locked to the fixed magnetic field, thereby stopping the motor.

DC link Also called a DC filter or DC bus. A filter circuit composed of an inductor and a capacitor. Used to filter or smooth the AC ripple from the output of the converter stage.

DC motor A device that converts electrical energy in the form of Direct Current into mechanical energy. A DC motor typically contains an armature, one or more field windings, a commutator, and brushes. Some DC motors have permanent magnets in place of armature windings. In this type of motor, called a brushless DC motor of BDCM, brushes are unnecessary.

digital input select If selected, this value bypasses the ramp time down setting, and the inverter decelerates the motor in the shortest possible amount of time as a result.

diode A two-terminal, solid-state device that allows current flow in one direction only. The terminals are the cathode and the anode. Sometimes called a p-n junction diode.

duty cycle The ratio of the time of the "on" pulse to the time of one complete cycle. A fifty-percent duty cycle has an "on" pulse duration one-half as long as one cycle.

dynamic braking A type of electrical braking in which the motor is disconnected from the power supply. The motor acts as a generator as it slows, and the motor's energy is dissipated through separate resistors.

efficiency When referring to motor efficiency, this is the ratio of output power to input power. Output power is typically expressed in horsepower and must be converted into watts (1 hp = 746 W). Input power is typically measured in watts. Therefore the formula for the ratio is: efficiency = $watts_{out}/watts_{in}$. The result is then multiplied by 100 and expressed as a percent. The higher the percentage, the greater is the motor's efficiency.

electronic variable speed drives Electronic circuitry used to control the speed, torque, and starting characteristics of a motor.

encoder An optical or magnetic device attached to the shaft of a motor to sense the motor shaft speed and/or position by generating voltage pulses. The voltage pulses are produced by a shutter that interrupts light striking an optical sensor such as an optocoupler or a phototransistor. The shutter rotates because it is attached to the shaft of the motor, which is in motion. In a magnetic encoder, the voltage pulses are produced when a magnet attached to the motor shaft revolves past a magnetic sensor such as a Hall-effect device.

error signal The difference between the command or reference signal and the feedback signal. The error signal is used by the drive to compensate for any variations in motor performance from the desired results.

excitation current DC current that produces the magnetic field in the field windings of a DC motor or generator. Excitation current can also produce the magnetic field in the rotor of an AC synchronous motor or in an alternator.

feedback signal A signal indicating motor performance. This signal can originate from the motor itself or from an external device mounted on the motor's shaft. The feedback signal is added to the command or reference signal to determine the error signal.

field Usually, the stationary portion of a motor or generator. The field typically contains windings.

four-quadrant chopper A chopper that operates in all four quadrants of motor operation: motoring forward, braking forward, motoring reverse, and braking reverse.

free-wheeling diode A diode connected across an inductor that is operating with DC applied. The free-wheeling diode provides a path for the field to collapse and discharge without harming other components in the circuit.

frequency As applied to drives, the output frequency of the inverter.

Hall-effect device A device that produces a voltage in the presence of a magnetic field.

IGBT Abbreviation for insulated gate bipolar transistor.

insulated gate bipolar transistor A voltage-controlled device with three terminals: the emitter, the collector, and the gate. The insulated gate bipolar transistor, or IGBT, is capable of high-speed switching at high voltages.

inverter A device that changes DC power into AC power.

jogging speed As applied to drives, sets the speed of the motor when jogging.

local/remote The inverter can be set for local control from the inverter's control panel or remote control from a start/stop station located away from the inverter cabinet.

magnetic encoder An encoder that uses a device such as a Hall-effect device to sense the motor shaft's position and/or speed.

manual control A control system in which motor performance is not sensed electronically and therefore can vary under changing load conditions. Because of this variability, motor performance must be monitored by personnel and corrective action taken manually. Also known as open-loop control.

maximum speed As applies to drives, depending on the setting, it may be possible to attain a speed higher than the rated speed of the motor. This parameter must be set higher than the minimum speed setting. If it is set lower than the minimum speed setting, the motor will not run.

megger Nickname for the megohmmeter.

megohmmeter A special type of ohmmeter that uses a high voltage to measure very high values of resistance.

metal oxide varistor A device used to provide surge protection. Also called an MOV.

minimum speed As applies to drives, the slowest speed setting at which the motor will run.

MOSFET A three-terminal, solid-state device. The terminals are: the gate, the source, and the drain. The gate is electrically insulated from the rest of the device. MOSFETs are used as electronic switches. The acronym "MOSFET" stands for *Metal Oxide Semiconductor Field Effect Transistor*.

motor magnetization Set to the no load current rating on the motor's nameplate.

motor nominal current Set to the full load current rating on the motor's nameplate.

motor nominal frequency   The rated frequency of the motor from the motor nameplate. This value should be set as closely as possible to the specified value.

motor nominal voltage   The rated line voltage of the motor from the motor nameplate. This value should be set as closely as possible to the specified value.

motor power   This is the motor's power rating expressed either in horsepower (HP) or in kilowatts (kw). Some drives will accept the value in either unit of measurement while others require converting the units from one measure to the other.

MOV   A device used to provide surge protection. Also called a metal oxide varistor.

NPN transistor   A transistor represented by an arrowhead pointing outward, away from the base terminal, on the emitter terminal. This transistor must be biased so that the base has more positive potential than the emitter.

one-quadrant drive   A drive that allows the motor to operate only in one direction and does not provide braking.

open-loop control   A control system in which motor performance is not sensed electronically and therefore can vary under changing load conditions. Because of this variability, motor performance must be monitored by personnel and corrective action taken manually. Also known as manual control.

optical encoder   An encoder that uses a device such as a photo-transistor or optocoupler to sense a motor shaft's position and/or speed.

overhauling   A condition that can occur when a motor is driving a load with high inertia. The load will have sufficient momentum to continue to rotate even after the motor has been disconnected from the power supply. An example of this phenomenon is a flywheel.

phase voltage   The voltage measured from phase to phase.

phototransistor   A transistor (npn or pnp) that uses a "window" in the package to allow light to strike its base. When light strikes the base of the phototransistor, the collector current increases.

plugging A form of braking whereby the motor, while running in one direction, is made to rotate in the opposite direction. Plugging is reserved for emergency braking.

P-N junction diode A two-terminal, solid-state device that allows current flow in one direction only. The terminals are the cathode and the anode. Sometimes called simply a diode.

PNP transistor A transistor represented by an arrowhead pointing inward, toward the base terminal, on the emitter terminal. This transistor must be biased so that the base has more negative potential than the emitter.

pulse frequency control Control pulses, applied to solid-state switches, control how often, or the frequency at which a device is turned on or off.

pulse generator A type of integrated circuit, used to provide a trigger pulse to control the output of the comparator stage.

pulse width control Control pulses, applied to solid-state switches, control the length of time a device is turned on or off.

pulse width modulation A technique used by electronic variable speed drives to vary the frequency of the voltage applied to a motor by varying the pulse width of the applied voltage. Also known as PWM.

PWM Abbreviation for pulse width modulation.

ramping A form of braking in which the current, voltage, and/or frequency supplied to the motor is controlled in a rapid or gradual fashion.

ramp time down As applied to drives, this is the deceleration time (the time required to get from maximum speed to minimum speed) expressed in seconds. If this time is set too short, the inverter can trip out.

ramp time up As applied to drives, this is the acceleration time (time required to get from minimum speed to maximum speed) expressed in seconds. If this time is too short, the inverter can trip out.

reference signal The level set of programmed into a drive to maintain the desired motor operating speed. Also known as the command signal or set point.

regenerative braking A type of electric braking in which the energy generated by the motor is returned to the power supply.

regulation A measure of how well a motor maintains its speed under varying load conditions. Regulation is determined by subtracting the full-load speed of the motor from the no-load speed of the motor. This difference is then divided by the full-load speed of the motor, and the result is multiplied by 100 and expressed as a percent. The formula is: %Regulation = $((Speed_{NL} \times Speed_{FL}/Speed_{FL}) \times 100$. The smaller the percent regulation, the better is the speed control of the motor.

relay output select   Provides a contact closure when the inverter is placed in the run mode.

SCR   A three-terminal, solid-state device that conducts current when properly biased and triggered. The three terminals are: the cathode, the anode, and the gate or trigger. An SCR allows current flow in one direction only. SCR is an acronym for *Silicon Controlled Rectifier*. Sometimes SCRs are called thyristors.

self-excited DC motor   A DC motor whose DC excitation current originates from the self-generated DC of the motor.

separately excited DC motor   A DC motor whose DC excitation current originates from a separate source of DC, such as a power supply.

series field   Windings in the field of a DC motor consisting of several turns of large-gauge wire. The series field is connected in series with the armature of the DC motor.

set point   The voltage level set or programmed into a drive to maintain the desired motor operating speed. Also known as the command signal or reference signal.

shunt field   Windings in the field of a DC motor consisting of many turns of small-gauge wire. The shunt field is connected in parallel with the armature of the DC motor.

silicon controlled rectifier   A three-terminal, solid-state device that conducts current when properly biased and triggered. The three terminals are: the cathode, the anode, and the gate or trigger. A silicon controlled rectifier, also called SCR or thyristor, allows current flow in one direction only.

slip compensation   Typically the factory setting for this value should be adequate. This setting is affected by the motor power, motor nominal voltage, and motor nominal frequency values.

start compensation   Typically the factory setting for this value should be adequate. This setting is affected by the motor power, motor nominal voltage, and motor nominal frequency values.

start/stop mode This value is programmed for the various types of start/stop circuits used. For example, a two-wire start/stop, a three-wire start/stop, a three-wire start/stop with a jog, and so on.

start voltage Typically the factory setting for this value should be adequate. This setting is affected by the motor power, motor nominal voltage, and motor nominal frequency values.

step-down chopper A chopper circuit in which the output voltage is lower than the input voltage. Sometimes called a buck chopper.

step-up chopper A chopper circuit in which the output voltage is higher than the input voltage. Sometimes called a boost chopper.

summing point When used in a feedback circuit, the junction where the algebraic sum of two or more signals is obtained.

switching amplifier field current controller A type of DC drive which switches the current to the field circuit on and off as indicated by the feedback device monitoring the motor's performance.

tachometer-generator A device mounted to the shaft of a motor. The tachometer-generator produces a DC voltage proportional to the speed of the motor. The generated voltage produced by the tachometer-generator is used as the feedback signal to control the motor's speed.

thermal motor protect Depending on the setting chosen, this parameter will either flash the display when the motor's critical temperature is reached or trip the inverter.

torque Turning or rotating force produced by a motor.

transistor A three-terminal, solid-state device. The three terminals are: the emitter, the base, and the collector. A transistor is a current-controlled device that can be used either as a switch or as an amplifier. Transistors are available in two types: npn or pnp. Both types operate identically, except that the flow of current is reversed. Transistors are often referred to as BJTs or bipolar junction transistors.

trip reset mode If selected, this parameter will prevent the inverter from restarting automatically after a trip.

two-quadrant chopper A chopper that operates in only two quadrants, for example, motoring forward and motoring reverse.

uninterruptible power supply A power supply that provides emergency power in the event of a main power supply failure. The transfer from the main supply to the uninterruptible power supply is automatic and immediate. Also know as a UPS.

UPS   An uninterruptible power supply.

variable voltage inverter   Also called a VVI. A type of AC drive in which the DC voltage is controlled, and the DC current is free to respond to the motor needs. A converter, DC link, and an inverter are used to vary the frequency of the applied AC voltage.

V/f ratio   Volts-to-frequency ratio. Also known as volts-to-hertz ratio or V/Hz.

V/Hz   The ratio of output voltage to frequency. This ratio must remain constant for the output torque to remain constant. If the output voltage is maintained while the frequency increases, the output horsepower remains constant, but the output torque decreases. Also known as volts-to-frequency ratio or V/f.

voltage   As applied to drives, the output voltage of the inverter.

voltage fed inverter   An inverter in which the DC bus voltage is controlled while the DC bus current varies to meet the needs of the motor. Also known as a VSI or voltage source inverter.

voltage source inverter   An inverter in which the DC bus voltage is controlled while the DC bus current varies to meet the needs of the motor. Also known as a VSI or voltage fed inverter.

VSI   Voltage source inverter. An inverter in which the DC bus voltage is controlled while the DC bus current varies to meet the needs of the motor. Also known as a voltage fed inverter.

VVI   Also called a variable voltage inverter. A type of AC drive in which the DC voltage is controlled, and the DC current is free to respond to the motor needs. A converter, DC link, and an inverter are used to vary the frequency of the applied AC voltage.

Index

Page numbers in *italics* indicate figures